P9-CKY-607

displacing place

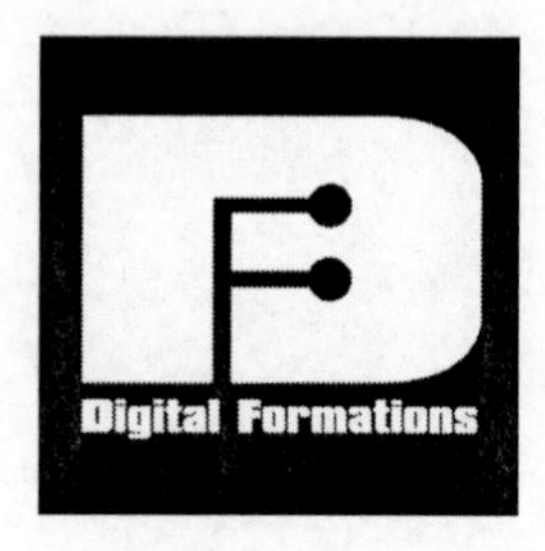

Steve Jones
General Editor

Vol. 42

PETER LANG
New York • Washington, D.C./Baltimore • Bern
Frankfurt am Main • Berlin • Brussels • Vienna • Oxford

displacing place

Mobile Communication in the Twenty-first Century

EDITED BY Sharon Kleinman

PETER LANG
New York • Washington, D.C./Baltimore • Bern
Frankfurt am Main • Berlin • Brussels • Vienna • Oxford

Library of Congress Cataloging-in-Publication Data

Displacing place: mobile communication in the twenty-first century /
Edited by Sharon Kleinman.
p. cm. — (Digital formations; v. 42)
Includes bibliographical references and index.
1. Wireless communication systems. 2. Telecommunication—Social aspects.
3. Mobile communication systems. 4. Space and time. I. Title.
TK5103.2.K588 303.48'33—dc22 2006101466
ISBN 978-0-8204-8660-4 (hardcover)
ISBN 978-0-8204-8659-8 (paperback)
ISSN 1526-3169

Bibliographic information published by **Die Deutsche Bibliothek**.
Die Deutsche Bibliothek lists this publication in the "Deutsche
Nationalbibliografie"; detailed bibliographic data is available
on the Internet at http://dnb.ddb.de/.

Cover design by Jean Galli

The paper in this book meets the guidelines for permanence and durability
of the Committee on Production Guidelines for Book Longevity
of the Council of Library Resources.

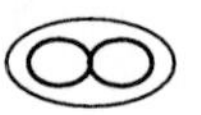

29 Broadway, 18th floor, New York, NY 10006
www.peterlang.com

Printed in the United States of America

Dedicated to Elaine and Arthur Kleinman,
and Bernice and Bertram Bradley,
for their boundless love and inspiration

CONTENTS

Acknowledgments ix

Preface xi

Introduction 1
Sharon Kleinman

PART ONE: PLACE AND 'POLIS

1 Mobile Communication in the Twenty-first Century or "Everybody, Everywhere, at Any Time" 7
Gary Gumpert and Susan J. Drucker

2 Municipal Wi-Fi Comes to Town 21
Harvey Jassem

3 Mobility in Mediapolis: Will Cities Be Displaced, Replaced, or Disappear? 39
Gene Burd

4 Living and Loving in the Metro/Electro Polis: Understanding the Neurobiology of Attachments in a Society with Ubiquitous Mobile Information and Communication Technologies 59
Yvonne Houy

5 Displacing Place with Obsolete Information and Communication Technologies 77
Julie Newman

PART TWO: MOBILE INNOVATIONS

6 Cyber-crime on the Move 91
Matthew Williams

7 Breaking Free: The Shaping and Resisting of Mobility in Personal Information and Communication Technologies 105
Julian Kilker

8 Mobile Culture: Podcasting as Public Media 123
Jarice Hanson and Bryan Baldwin

9 Reach out and Download Something: An Analysis of Cell Phone and Cell Phone Plan Advertisements 141
Richard Olsen

PART THREE: MOBILE TECHNOLOGIES AT WORK

10 Networks Unleashed: Mobile Communication and the Evolution of Networked Organizations 159
Calvert Jones and Patricia Wallace

11 Medical Communication: Improving Patient Safety in the Operating Room and Critical Care Unit 175
Keith J. Ruskin

12 Therapy at a Distance: Information and Communication Technologies and Mental Health 189
Penny A. Leisring

13 But You Don't Play with the Mobile Information and Communication Technologies You Already Have: An Instructional Technologist's View of Teaching with Technology in Higher Education 207
Gary Pandolfi

14 Pumping up the Pace: The Wireless Newsroom 215
Andrew Smith

Conclusion: Anytime, Any Place: Mobile Information and Communication Technologies in the Culture of Efficiency 225
Sharon Kleinman

About the Contributors 235

Index 241

ACKNOWLEDGMENTS

It takes a village . . .

First, I thank the chapter authors for their extraordinary contributions and collaboration.

I am grateful to Steve Jones, Sophie Appel, Mary Savigar, Damon Zucca, and the outstanding staff at Peter Lang for their guidance and support for this book.

I also appreciate the tremendous support from my colleagues at Quinnipiac University. Special thanks to David Donnelly, Linda Broker, Kathleen McCourt, John Lahey, Cynthia Gallatin, John Paton, Bill Clyde, Michele A. Moore, Rick Hancock, Raymond Foery, Rich Hanley, John Gourlie, Grace Levine, Mira Binford, Margarita Diaz, Paul Janensch, Karin Schwanbeck, Vicki Todd, Frances Rowe, Lauren Erardi, Robin D'Errico, John W. Morgan, Janet Waldman, Janice Ammons, Charles Brooks, Angela Mattie, Jonathan Blake, Liam O'Brien, Carol Marro, Angela Bird, Barbara Baker, Aileen Moran, Carrie Bulger, Lynne Hodgson, Rachel Ranis, Michele Hoffnung, Janet Valeski, June DeGennaro, Frances Kelly, Robert Young, Sheila Pallato, Nancy Kapchan, Kathy Cooke, David Valone, and Jill Fehleison.

Jarice Hanson encouraged me to embark on a book project and provided sage advice about academic publishing for which I am grateful.

Michelle Komie imparted helpful guidance on the proposal.

Ti Badri, my graduate research assistant, provided superb research and editorial assistance.

Stacey Donovan, Ellis Perlswig, M.D., Jenna Jambeck, Joshua Kim, Robert Miller, and the Hon. Herbert Smolen reviewed chapters and generously shared their technical expertise.

Julian Kilker suggested *Displacing Place* for the title of the book.

Jean Galli masterfully designed a provocative and fun cover.

The staff and regulars at Lulu's and Koffee—New Haven cafés that I affectionately call my satellite offices—supplied cappuccinos and conversations that aided the writing and editing processes.

My students shared wonderful insights about life in the twenty-first century with candor and optimism.

I am forever grateful to my friends Colleen Sexton, James Williams, Katie Tranzillo, Theo Forbath, Diane Bauer, Nita Maihle, Jerry Saladyga, Daniel Pardy, Cam and Donna Staples, Mary Mayer, Victoria Jones, Alex Heonis, Jennifer Jacobs, Joan Bombalicki, and Dena Miller, M.D., and my family, Elaine and Arthur Kleinman, and Bernice and Bertram Bradley, for their love and rock-solid support.

Finally, for inspiration and camaraderie, I thank the New Haven yoga community. Special thanks to Maureen McGuire, Heidi Sormaz, Lillee Chandra, Kirsten Collins, and Peg Oliveira. Namaste.

PREFACE

Alex's silver cell phone lay on a shelf next to a jar of scissors and combs. When the phone emitted a soothing piano arpeggio, she stopped cutting my hair to check the caller ID. After confirming that it wasn't someone calling from her six-year-old daughter's school, she picked up her shears again. I asked Alex how people managed before we were accessible virtually anytime and any place, and we both laughed. Then she asked how this book was inspired.

Displacing Place grew out of conversations with my mother about technological innovations during her lifetime. Her mother's mobility had been severely limited during the last few years of her life due to illness. My mother remarked that my grandmother would have benefited tremendously from technologies that we take for granted today, such as cordless phones, voicemail, and television remote controls. I remembered that when we telephoned my grandmother, we would wait about fifteen rings for her to answer, because it took her that long to get up and walk to the one phone in her apartment, which was mounted on a wall in the dining room. It must have been profoundly frustrating for my grandmother when she walked as quickly as she could to answer the phone and the caller hung up before she reached it. Without an answering machine, voicemail, or caller ID—they hadn't been invented yet—she had no way of knowing whose call she had missed.

In contrast to my grandparents' era, mobile communication and nearly seamless anytime, any place connectivity characterize the twenty-first century. Worldwide, more and more people are using portable devices for communication, information seeking, and entertainment. In unanticipated and far-reaching ways, these technologies are impacting how and where people work, play, and relate to one another. The essays in this collection explore the nuances of these transformations.

Sharon Kleinman
New Haven, Connecticut

INTRODUCTION

Sharon Kleinman

While I was walking in the woods near my home on a Saturday afternoon, my friend James called me on the cell phone from Best Video, a local independent DVD store, to invite me over to his home for dinner and a movie. "Have you seen this one?" he inquired, before selecting a title to rent. Later that night, I responded to questions from my students via email, a five-pound computer propped on my lap while I sat in bed. Meanwhile, a few houses away, my radiologist friend Dena was in her home office reading images of the insides of human bodies generated by magnetic resonance imaging (MRI) and computed axial tomography (CAT) machines located in a different state. Twenty-five New York City–area hospitals transmitted images for diagnosis to her and a cohort of radiologists scattered around the country. Teleradiology is a perfect example of "displacing place," she noted when I told her about this book. She remarked that she enjoyed being "at work" from home, reading MRIs and CAT scans and bantering via instant messenger with other doctors.

A defining characteristic of life in the early twenty-first century is the ubiquity of mobile communication. Mobile information and communication technologies (ICTs) are enabling people to participate in new ways and in additional contexts in a broad range of activities. This book aims to add to our understanding of the implications of mobile ICTs with essays written by leading-edge scholars and professionals from a variety of fields that explore some of the many stories of mobile communication.

The book's title, *Displacing Place*, refers to the circumstances of mobile communication: "here" and "there" can be virtually anywhere, and, moreover, both can be moving. Key features of mobile ICTs are their portability and their capacity for enabling people to communicate, seek and share information, and be entertained in ways that transcend spatial and temporal constraints.

Displacing Place is divided into three parts. In part 1, "Place and 'Polis," the first five chapters focus on the wide-ranging social implications of mobile communication, exploring emerging issues and looking ahead to the future. In part 2, "Mobile Innovations," the following four chapters reveal how producers and users are imaginatively transforming the features, functions, and symbolic meanings of these technologies. In part 3, "Mobile Technologies at Work," the next five chapters examine how mobile devices are being "put to work" to enhance professional effectiveness in a range of fields.

More specifically, in part 1, Gary Gumpert and Susan J. Drucker set the stage, introducing a new paradigm for mobile communication and analyzing how mobile ICTs influence relationships to place.

Harvey Jassem looks at wireless fidelity (Wi-Fi) as part of the broadband information highway, examining the roles governments play in its development and the impacts of universal Wi-Fi access on community life.

Gene Burd considers why mobile ICTs are the latest concern of those predicting the downfall of the city. He argues that mobile communication is modifying cities, uniting the virtual and real, forcing old media to merge and readjust, which will ultimately lead to the rediscovery of interpersonal communication inherent in the nature of cities.

Yvonne Houy explores the implications of findings in the emerging field of interpersonal neurobiology about the formation and functions of human attachments for a society that increasingly lives in a hybrid physical and virtual space through mobile ICTs.

Julie Newman scrutinizes the inherently limited life cycle of ICTs and exposes the environmental and human health impacts of their improper disposal, adding another dimension to the discussion of the effects of ICTs on place.

In part 2, Matthew Williams unveils new cyber-criminal threats facilitated by mobile ICTs, ranging from school bullies harassing fellow students via text and multimedia messaging beyond the school's physical and temporal boundaries, which threatens the sanctity of the home, to hackers viewing the contents of mobile phones, which threatens individual privacy and organizational security.

Julian Kilker focuses on how people inventively modify the characteristics of mobile phones, game devices, and media players to expand where, when, and how media are consumed.

Jarice Hanson and Bryan Baldwin evaluate whether podcasts represent a new generation of public media that democratically foster a diversity of viewpoints and serve the public interest.

Richard Olsen analyzes how cell phones are defined and promoted to potential users through advertisements for cell phones and cell phone plans. He highlights the shift from ads emphasizing connection to those emphasizing stimulation, and he elucidates how advertisements inform as well as reflect people's perceptions and usage of mobile ICTs.

In part 3, Calvert Jones and Patricia Wallace show how mobile communication is facilitating the processes and progress of networked organizations, transforming workplaces, and affecting individuals' work-leisure balance.

Keith J. Ruskin looks at medical communication and assesses how new communication technologies coupled with the easing of hospital restrictions regarding their usage will improve patient care in the operating room and critical care unit.

Penny A. Leisring discusses telehealth approaches within clinical psychology, revealing how the Internet is being used to extend psychological services to people who might not otherwise seek treatment and how Web-based adjunctive services are being used to enhance traditional therapeutic interventions.

Gary Pandolfi evaluates the integration of ICTs into higher education, providing scenarios that illustrate when a new technology might improve teaching and learning, and when it merely produces a new look for the status quo.

Andrew Smith conveys how cell phones, laptop computers, and personal digital assistants (PDAs) have transformed journalism, making it easier for editors to locate reporters and to reduce reporting time, enabling news organizations to get stories on the air, onto the Web, and into newspaper pages faster.

In the conclusion, I reflect on how mobile ICTs are changing our communication opportunities and expectations and altering the nature of communication in public places. I also call attention to situations in which people are regulating or eschewing anytime, any place connectivity in order to decelerate the pace of life and focus rather than multitask, which raises questions about how the stories of mobile communication will unfold in the years to come.

Together, the essays in this collection convey new insights about the impacts, symbolic meanings, and future of mobile ICTs that I hope will be valuable not only to students and researchers, but to all readers interested in social and technological trends in the twenty-first century.

· 1 ·

PLACE AND 'POLIS

· 1 ·

MOBILE COMMUNICATION IN THE TWENTY-FIRST CENTURY OR "EVERYBODY, EVERYWHERE, AT ANY TIME"

Gary Gumpert and Susan J. Drucker

The Parable of the Mobile Rock: Displacing Place

In the beginning there was a rock. It was a lone rock. One day, for an unknown reason, the rock split and one became two and thus they discovered the need to communicate with each other. It is not easy for rocks to communicate.

There was a time when we lived relatively sedentary lives with safety, security, and obligation defined in terms of place. It was left to the adventurers, the exiled, and nomads to dare venture forth into distant places. That was not long ago. Less than two hundred years ago communication was either face-to-face or written in the form of letters, journals, newspapers, and books. For the purists, smoke signals, carrier pigeons, and messages set adrift in bottles were readily available. The alternatives were limited.

In the twenty-first century, the cornucopia of communication devices makes it difficult to even define the simple nature of communication between two or more persons—or their representative machines. On a basic level, the mobility of humans can be defined in terms of transportation or by means of communication devices. In the first instance, we are physically transported from one destination to another in or on a moving vehicle. Alternatively, a representation of ourselves can be transported in a mediated form—through a medium of communication in which one or more senses is preserved for later perusal and/or sent elsewhere to a distant location.

COMMUNICATION	
Locomotion or Transportation of Person	Communication through Medium

Communication is further complicated by hybrid forms and situations that mix transportability and electronics in the form of person and device.

MEDIUM	
Physical Object: book, newspaper, letter, etc.	Electronic: radio, television, telephone, etc.

There is a sense of mobility inherent in all those instances of communication in which human interaction is extended beyond the site of the face-to-face act, as information or data in the form of print, voice, and visual is sent elsewhere. Mobility, in regard to communication, however, has taken on another sense of meaning. It is taken for granted that we are able to communicate from one site to another, but "the moving site" represents the convergence and transformation of communication technology into a nonplace event.

Over forty years ago, Melvin M. Webber (1964) described the transformation of community from propinquitous to nonpropinquitous, from a place to a nonplace orientation, as spatial proximity or nearness no longer became a necessary prerequisite for the relationships of social community. "With few exceptions, the adult American is increasingly able to maintain selected contacts with others on an interest basis, over increasingly great distances and is thus a member of interest-communities that are not territorially defined" (p. 111). More recently, Webber (1996) suggests that "the metropolis is a massive communications switchboard. . . . It exists only because interdependent persons and groups have to be accessible to each other and because the cost of overcoming space has not yet reached zero." Hopefully we shall never arrive at that state, but the global saturation of the environment with portable communication devices is well on its way.

This chapter attempts to define the broad parameters of mobile communication both in terms of technology and in regard to the social-communal aspects of the phenomenon. In particular, it examines the connections among mobility, place, and community. A number of questions are raised and some are left unanswered in reference to the exuberant adoption of communication innovations. What is community in a technological age? Is it possible to retain vibrant place-based communities while recognizing the absolute dependence upon digitalization and the increasing significance of digital communities? What do the burgeoning capabilities of Internet, mobile, and wireless communications mean for our concepts and experiences of public and private spaces in the city? How might policies and practices of urban design and planning best address the changing nature and experience of urban space shaped by mobile media?

One estimate made regarding the global sales of mobile phones is that sales will exceed 1 billion units by 2009 ("Gartner," 2005). With the addition of Wi-Fi accessibility, laptop computers now account for over 50 percent of all computer sales ("U.S. Laptop Computer Sales," 2005). In addition, the market has seen a surge in the sale of smart phones.

The New Media Landscape

The "mass communications switchboard" is an antiquated metaphor, a manually operated apparatus for interconnecting telephone lines and routing telephone calls. But the Webber addendum is fascinating because it suggests a point in time when place is irrelevant and mobility is an expectation. An updated paradigm of mobile communication would include a communication device that:

1. Is potentially multifunctional, serving a number of purposes;
2. Is multimodal in that it has the potential of utilizing more than one means of transporting data and utilizes more than one-sense modality;
3. Facilitates flexibility of destination in terms of scope and size of audience—narrow, broad, mass, blog, and so forth; and
4. Is capable of connecting one or more receivers, but where location is irrelevant to connection.

Thus, for example, a laptop computer is a portable device that allows for the transmission of print, voice, recorded audio, and picture. A mobile telephone allows for the transmission of data, voice, and text messaging. A smart phone is a handheld device that includes a mobile phone, a personal digital assistant (PDA), and other assorted information appliances.

It is this multifacetedness that is accompanied by expectations shaped by the intrinsic properties that constitute and define each medium:

1. Interactivity/interaction: The action of the user generates a response either from another human being or from a computer program. Interactivity is associated with action-reaction or a command-response exchange. Interactivity allows a user to make choices that result in a variety of responses.
2. Immediacy of response: This refers to the relative ease with which a sender receives a response to a specific message. Swift feedback becomes an expectation, a norm. The expectation of immediate connection and prompt response results in willing interaction with either a constructed artificial response system or a human being—sometimes they are indistinguishable.
3. Increased sense of control: The choice of contact, channels, and time is an essential component and expectation. Digital media offer "bidirectionality." Data can flow in both directions. The increased ability to "block" or "pull" in data or programming promotes a sense of control.
4. Immediacy of connection: Space and location are less significant if not irrelevant. Connection is the primary goal of a digital environment. Specific locality becomes merged with global access. Local becomes global and global local.

Global Mobile Communicators

There is yet another sense of mobility in which communication device, person, and transportability are fused and in which the distinctions among the three are unclear, if not ambiguous. Person becomes object when subjected to a global positioning system (GPS)—a satellite-based system consisting of a network of twenty-four GPS satellites that orbit the Earth twice a day and transmit signal information. GPS receivers take this information and use triangulation to calculate the user's exact location. What we have here is a very strange, almost contradictory condition. We suggested earlier that mobile communication is intimately linked to the disconnection of person from place, of the weakening of neighborhood or community ties as more appealing, self-selected non-place communities are accessible. The mobile emancipation of person and physical location is paradoxical. The flip side of delocation is the ability to locate. As we increase our ability to communicate to any place from anywhere at any time, we are subject to pinpoint location by ourselves or others as we move. We require global positioning to locate the mobile "us" in physical space.

Mobile communication can be defined technologically, but because each technological communication device has residual societal consequences, the social implications are critical and probably not apparent.

The technology of mobile communication is the result of digitalization, miniaturization, and protocol. The first two characteristics are self-evident; the latter refers to a universal agreement as to how digital information will be transported from device to device. Without those three characteristics, convergence and multifunctionality would not be possible. Digitalization alters and irrevocably modifies the nature of relationships between individuals and their environment (Gumpert & Drucker, 2005).

Physical Space, Mobility, and Media Space

Considerable research is available on the impact of communication technology on the future form and shape of places. The research focuses on recombinant hybrid, smart, and wired and unwired cities. In addition, there is a significant body of literature growing around virtual, online, digital communities. The latter literature stresses the connection of individuals in an electronic public space transcending physicality of the environment. While the state of digital communities is important, our focus is on the symbiotic relationship between public space and mediated communication, on the impact of mediated communication upon the function and design of experiencing physical public spaces that traditionally have been vital to the formation and maintenance of community.

In a previous article, we introduced the concept of "displacement"—the "reciprocal and defining interdependence of place modified by communication technology" (Gumpert & Drucker, 2005, pp. 374–375). "Displacement" is a temporal concept, defined by a difference in time or time usage; the amount of time spent in media activity fundamentally alters the amount of time available for other events. It is thus self-evident that the contemporary individual spends an increasing amount of time electronically connected with others in lieu of physical interaction on a face-to-face basis. The process of mediated connection brings with it a degree of "replacement" as well as "displacement." "Replacement" is a spatial concept, a difference of environment, the substitution and/or alteration of one or more locations for another. Both displacement and replacement portend significant consequences for the nature and use of public space. We would now suggest "a-location" as a consequence of media mobility. A-location refers to the redefining of social space and psychological presence with its potential emancipation from physical place.

Psychological presence in physical space is altered through ubiquitous, flexible, and mobile media connection. Presence refers to the illusion of being there, whether "there" exists in the physical space or not (Biocca, 1997). As mobile connection beckons, people can become immersed or absorbed in media connection, altering their awareness and interaction with the physical environment.

At the heart of the matter is the essential role of the human communicator gliding between physical and media environments. It is this sense of contraction that permeates the relationship of person and place when a medium of communication is interposed. Indeed, the whole person-media experience is a bit perplexing as

> the more we extend our connection, the more insular we become. The more we control our communication environment, the less is surprise or chance a daily expectation. The more we connect, the more we seek to control the connection. The more we detach from our immediate surroundings, the more we rely upon surveillance of the environment. The more communication choice offered, the less we trust the information we receive. The more information and data available, the more we need. The more individuality we achieve, the more communities we seek. The more we extend our senses, the less we depend upon our sensorium. (Gumpert, 1996, p. 41)

Timo Kopomaa (2004), a Finnish urban sociologist, argues that mobile phones can be viewed as a "'place' adjacent to, yet outside of home and the workplace, a third place in the definition of Ray Oldenburg (1989)" (p. 269). Oldenburg defined his concept in terms of physical spaces, seen as neutral ground. Kopomaa argues that "modern forms of public engagement and urban social interaction include mobile togetherness, where people meet each other on the move, and stationary togetherness, where people gather in places, such as waiting halls or public transportation" (p. 271). Kopomaa contends that "digital life thrives in public places," but the use of technologies like the mobile phone implies a privatization of public space (p. 271). "It involves a relationship with the milieu, which implies a yearning for privacy and a place for oneself" (p. 270).

Portable Public Privacy

Public interaction is being transformed into "disembodied private space" by mobile technologies. Human beings have always constructed their own sense of space as they enter a public place. They have long erected "media walls." A newspaper, magazine, or book raises a barrier separating the reader from place and signals a withdrawal from potential face-to-face interaction. A wall of sound, the acoustical space that surrounds each of us, is constructed through silence or acoustical environments not shared with others. The portable headsets and earbuds transmitting radio programs, and cassette, CD, and MP3 players playing music, allow people to be mobile and private at the same time. The cassette-playing Walkman, introduced in 1979, became the CD player, which became the iPod, "the 21st century Walkman" introduced in 2001 by Apple Computer (Stross, 2005). Podcasting is a relatively new delivery system through which existing and unique new content and forms of programming are distributed (see Hanson & Baldwin, chapter 8 in this volume). Short-form original content (mobisodes) for mobile phones is composed with a two- to three-inch screen in mind.

The private sound (and video) experience excludes others and deters potential interaction in public space. Mobile telephony, and to a lesser extent PDAs and Wi-Fi Internet access, are creating even newer media walls. In *The Cybercities Reader*, Stephen Graham (2004) observes:

> Within a short time mobile communication technologies have transformed the nature of city street life due to private electronic interchanges. . . . Mobile and personal digital assistants (PDAs) are key interfaces through which many urban residents shape and experience city life. . . . Saturation of the city with mobile phones and other personal mobile ICT technologies heralds a reconstruction of the way city spaces are used, appropriated and mediated. This changes public choreography of physical movement in the city. (p. 133)

Portable public privacy meets the needs of people alienated and disconnected from and distrusting of public space. Not only does the mobile phone privatize public space, but its use has begun to alter our acoustical expectations. Some media walls are more permeable than others.

Acoustical Space

The opposite of portable public privacy walls can be seen in the long-recognized phenomenon of auditory aggression. In *The Tuning of the World*,

R. Murray Schafer (1981) notes that "when sound power is sufficient to create a large acoustic profile, we may speak of it, too, as imperialistic" (p. 214). The linking of territory and sound is central to understanding attitudes toward media use. Spatial relationships are altered as the individual adapts to changes in personal space. A body of literature exists on the study of proxemics and the personal use of space. Vocal decorum is related to social distance (Gumpert, 1987). Personal space surrounding an individual is measured visually and acoustically. The concept of acoustical space and its relation to territoriality is not new. Schafer suggests that the acoustic space of a sounding object is that volume of space in which the sound can be heard. The maximum acoustic space inhabited by an individual will be the area over which his or her voice can be heard. The acoustic space of a radio will be the volume of space in which those sounds can be heard. Schafer (1981) concludes, "modern technology has given each individual the tools to activate more acoustic space" (p. 214). Mobile telephone conversations are often seen as forms of vocal aggression, as private mediated conversation intrudes on more usual and acceptable forms of behavior.

Responses to the problems of acoustical space violations have taken many forms, including regulatory responses, such as creating mobile phone-free environments in restaurants, schools, hospitals, and theaters, or providing designated areas for use in aptly named commuter train "babble cars."

Publicness and Privateness in Public Space

Public space is where community rituals are enacted and role identity performed. Rituals of public behavior include the tacit acknowledgment of others: greetings, nods, handshakes, and kisses. They are the phatic elements that serve to facilitate potential meaningful interaction. In today's world, such rituals conflict with competing rituals of mediated behavior. These rituals are superimposed upon the rituals of public behavior.

For centuries, participation in public life has been an essential component of community. Public space, a manifestation of community, functioned as a medium through which community was sustained. Public spaces are marked by boundaries; they are defined by measurable parameters. Private connection and public life illustrate the difficulty in fixing boundaries of publicness and privateness. The "wired" individual entering public space is physically located in immediate surroundings and simultaneously disconnected from that physical environment. What we have is psychological disconnection prompted by media connection. Public spaces provide interactional potential, contacts both welcome and unwelcome. Psychological presence, a state of subjective perception, is filtered through the mediated experience. Presence, the conscious state of awareness and attendance, shifts back and forth from the physical to the media space.

The people of the new media generation, tutored on digital media, are marked by characteristics of digital media: multitasking talents and the expectation of immediate connection and prompt response, control of personal contact and information, access regardless of distance. They operate in a new psychosocial state spanning and linking the physical and mediated environments.

Mobility as Control

In 2006, a controversy erupted in New York City when the schools chancellor and the mayor imposed a ban on cell phones within the school grounds.

> But Mayor Michael Bloomberg and Schools Chancellor Joel Klein have staunchly refused to drop the ban. They insist cell phones are a distraction and are used to cheat, take inappropriate photos in bathrooms, and organize gang rendezvous. They are also a top stolen item.
>
> Students have refused to give up their phones, saying the devices have become too vital to their daily existence and to their parents' peace of mind.
>
> "My mother, she needs me to have the cell to call me and check up on me," said Steven Cao, 16, a sophomore who lives in Staten Island and attends Stuyvesant High School in Manhattan. He called the ban stupid. ("School Cell Phone Ban," 2006)

Parents maintain their legal right to monitor their children through cell phone connection. Since the late 1980s, the New York school system has barred beepers and other communication devices from the classroom, but the policy has been transformed into an "out of sight, out of trouble" approach. But the disruptive impact of the cell phones in the classroom, with ringing phones, text messaging, photographing, cheating, and game playing, has become epidemic. In April 2006, city officials used random portable metal detectors at some schools to keep out weapons, but that process also led to the confiscation of hundreds of cell phones ("School Cell Phone Ban," 2006). The issue is fascinating because it introduces a surprising twist to the liberating quality of cell phone mobility.

The Anxiety of Disconnectedness

Many of us have experienced the pangs of discomfort that come with disconnection—the first day of school or summer camp come to mind as traumatic experiences for both parents and children. Yet realistically it was understood that temporary separation was inevitable, was part of maturation and growth. Mobile connection extends the trauma into everyday experience, whether a matter of control or a matter of anxiety. Businesses expect the constancy of contact. Husbands and wives need to know where the significant other is at any moment in time. We are insecure unless we know that we are reachable. Our expectations with regard to connection and disconnection have changed with the technologies of mobile media. Studies have been conducted tracking people's reactions to leaving a mobile phone at home. In Japan, respondents indicated feelings ranging from unease and feeling low to panic (Itoh, 2006).

Regulating the Displaced Place

There was a time when regulation of both spaces and media was based solely on geography. The regulation of broadcasting was based upon the public interest of a place-based community. There was also a time when the telephone defined community—defined it by operators, party lines, the name of a person or a business's telephone exchange, and later by the area code and dialing prefix. We are now in the process of "delocationizing"—triggered by the rise of the mobile media operating from "nowhere" and "anywhere." There was a time when the physical environment was regulated based upon distinct land usage, but that time may be changing as well. To the extent that the values, norms, and issues of any era are revealed by a society's laws, the threat posed by mobile media to traditional legal concepts is illuminating yet surprisingly neglected.

Zoning Mobility

Back in 1991, we looked at zoning as one of the most traditional regulatory devices shaping, perhaps determining, urban communication patterns. We began the zoning article in the following way: "Although interaction has been emancipated from place, public places still function as sites of face-to-face interaction. Today, spatial propinquity, the actual physical nearness of individuals, is less important to the maintenance of social communities than in the past because the modern media have eliminated 'cohabitation of a territorial place' as a prerequisite for community" (Drucker & Gumpert, 1991, p. 294). We noted that "zoning laws, through which design and planning decisions are implemented, serve as the vehicle for examining" connections between social interaction, media development, and environmental planning (p. 294). Zoning is the legislative tool that limits use of property (Mandelker & Cunningham, 1985). Concern for health and safety led to the partitioning of cities into clearly defined use districts. At that time, we examined the history of zoning laws as they regulated face-to-face interaction by restricting or stimulating communicative activity either explicitly or implicitly. The form of zoning we explored partitioned cities into areas based on function. The ability to zone by carrying out variations on residential, commercial, and industrial zones was predicated on the existence of both transportation and media systems that provide movement and communication between precincts.

In her classic book *The Death and Life of Great American Cities*, Jane Jacobs (1961) argued that livable cities require mixed-use neighborhoods. Livable neighborhoods require mixed uses; streets filled with surprises provide vibrant environments supportive of social interaction at street level.

In the intervening years, the concept of zoning has been the subject of several interesting new perspectives, most notably in the work of the New Urbanists, who stress the importance of the built environment in fostering community but emphasize that there must be a change to overcome the diminished sense of community and "civic deficits." They argue that the reorganization of physical space intimately is needed as a result of understanding the needs of citizens as social beings. The New Urbanists contend that suburban development creates social problems by undermining intergenerational social cohesion and community development, and they seek to integrate community-forming elements into planning of real neighborhoods (Grabill, 2003). The New Urbanists call for mixed-use zoning and multipurpose environments. They note that "streets, like land use, can no longer afford to be single purpose" (Calthorpe, 2002, p. 4). They urge the revision and rewriting of zoning codes that previously mandated separation of residential from commercial uses, low densities, and broad streets, while calling for zoning to mandate not only mixed uses but high densities.

As mobile technologies emancipate individuals from place, they simultaneously facilitate reentry into more diverse and often public spaces. "As Bill Mitchell argues, we will live in cities that are fluid mixtures of the real and the electronic suggesting that even though telecommunications allows information to be accessed from anywhere, the city itself does not cease to matter but rather remains vital in its electronically augmented form" (Johnson, 2004). Mitchell (2003) emphasizes that mobility fundamentally changes the way we use space and interact in our communities. The addition of Wi-Fi capability may indeed impact upon the design of buildings, plazas, parks, and entire cities as universal access becomes available. The superimposition of the digital infrastructure on the physical infrastructure increases the multifunctional nature of spaces. Calvin Johnson (2004), in his review of Mitchell's *ME++: The Cyborg Self and the Networked City*, notes:

> While 20th century city planning segregated different functions into separate zones, so that polluted industrial areas, for example, would not adjoin bucolic residential tracts, Mitchell anticipates that 21st century planning guidelines, while maintaining distance where appropriate, will accommodate several activities at any given location. Previous attempts at spatial multifunctionality—which, like sofa beds or mobile partitions, were often cumbersome—will go by the wayside as an adaptive, intelligent architecture permits, or even provides, the different software that users need for their PDAs or laptops in order to accomplish a range of tasks in a single space. . . . Furthermore, because networked systems are "fluid and amorphous," many of the activities happening at a particular place within the city will be unanticipated.

The call for mixed-use zoning takes on new meaning when one cannot anticipate specified use and when psychological presence in physical space can be so easily altered through ubiquitous media connection. Connectivity and interactivity on demand seamlessly accessed regardless of location creates a mixed-use zone of an unprecedented scale and nature.

The mixed-use zones of modern urban environments embody convergence of media, convergence of electronic and media space, convergence of digital and physical presence, and convergence of public and private domains.

Conclusions

The combination of the physical and virtual technologies of transportation and communication defines cities, suburbs, and rural areas, and the relationships among them. Cell phone towers, broadband, and Wi-Fi rollout challenge public and private policymakers who seek to find ways of maintaining, updating, and improving cities by integrating the latest technologies.

From a design perspective, there is a degree of tension that exists between the interactional potential of public (physical) space and that of an imposed sense of space. We could argue that the mobile telephone, for example, increases the number of people who psychologically can inhabit a space, but also decreases the number of people who can effectively communicate in that same space without creating noise. Thus, ten individuals speaking to someone else on a cell phone are different from ten people speaking to one another in the same room. If one were to design a room for mixed use (mobile and nonmobile phone users), would the designed space be the same? There is a new kind of interaction of people who are wired displacing a public place. Design challenges abound. Can public spaces be designed differently using different materials?

In summary, there are several key points to be made with regard to the relationship of mobile technologies, perception, and the use of physical public space. The social life of public space now competes with media technology that shifts interaction inward. Public interaction is being transformed into "disembodied private space" by mobile technologies. Not only does the mobile phone privatize public space, but its use has begun to alter the acoustical expectations of people in public spaces. Private connection and public life illustrate the difficulty in fixing boundaries of publicness and privateness. The rise of private space has been dominant, but now public space may once again draw people to those spaces that are rejuvenated through mobile communication technologies. An effectively designed new public space might complement physical environments and media alternatives. Ultimately, challenges abound for those concerned with communities and the regulation and design of places being transformed by mobile media.

Communication technologies are not reversible. One cannot undiscover media developments; one cannot reject the concept of mobility. It has become an integral part of our social values. The impact on the traditional relationships among place, person, and community is significant and needs consideration as we plan and design the fixed physical places being altered by mobile media.

References

Biocca, F. (1997). The cyborg's dilemma: Progressive embodiment in virtual environments. *Journal of Computer-Mediated Communication*, 3(2). Retrieved from http://www.ascusc.org/jcmc/v013/issue2/.

Calthorpe, P. (2002). The urban network: A new framework for growth. Calthorpe and Associates. Retrieved January 14, 2006, from http://www.calthorpe.com.

Drucker, S. J., & Gumpert, G. (1991). Public space and communication: The zoning of public interaction. *Communication Theory*, 1(4), 294–310.

Gartner says mobile phone sales will exceed one billion in 2009. (2005). Media Relations Press Release. Retrieved October 2, 2006, from www.gartner.com/press_releases/asset_132473_html.

Grabill, S. J. (2003, Spring). Markets and morality (The Acton Institute). *Charter of new urbanism*. Retrieved January 12, 2006, from http://www.cnu.org/about/index.cfm.

Graham, S. (Ed.). (2004). *The cybercities reader*. London: Routledge.

Gumpert, G. (1987). *Talking tombstones and other tales of the media age*. New York: Oxford University Press.

———. (1996). Communications and our sense of community: A planning agenda. *InterMedia*, 24(4), 41–44.

Gumpert, G., & Drucker, S. (2005). The perfections of sustainability and imperfections in the digital community: Paradoxes of connection and disconnection. In P. van den Besselaar & S. Koizumi (Eds.), *Digital cities III: Information technologies for social capital: Cross-cultural perspectives* (pp. 369–379). Berlin: Springer.

Itoh, S. (2006, September). The mobile phone's effect on everyday social/physical spaces, use of mobile phones and perception of space of Japanese university students. Paper presented at the International Association of People Environment Studies Conference, Alexandria, VA.

Jacobs, J. (1961). *The death and life of great American cities*. New York: Random House.

Johnson, C. (2004). Review: William J. Mitchell, *ME++: The cyborg self and the networked city*. *Technology and Cities*, 6. Retrieved October 11, 2006, from http://www.americancity.org/print_version.php.

Kopomaa, T. (2004). Speaking mobile: Intensified everyday life, condensed city. In S. Graham (Ed.), *The cybercities reader* (pp. 267–272). London: Routledge.

Mandelker, D., & Cunningham, R. (1985). *Planning and control of land development: Cases and materials* (2nd ed.). Charlottesville, VA: Michie.

Mitchell, W. J. (2003). *Me++: The cyborg self and the networked city*. Cambridge, MA: MIT Press.

School cell phone ban causes uproar. (2006, May 12). *CBS News*. Retrieved October 5, 2006, from http://www.cbsnews.com stories/2006/05/12/national/main1616330.shtml.

Schafer, R. M. (1981). *The tuning of the world*. Philadelphia: University of Pennsylvania Press.

Stross, R. (2005, March 13). How the iPod ran circles around the Walkman. *New York Times*. Retrieved from http://www.nytimes.com/2005/03/13/business/worldbusiness/ 13digi.html?ex=1268370000&en=b90493bfe6c9e003&ei=5090&partner=rssuserland.

U.S. laptop computer sales outstrip desktops. (2005). *Laptopical*. Retrieved October 3, 2006, from http://www.laptopical.com/laptop-compter-sales.html.

Webber, M. M. (1964). The urban place and the nonplace urban realm. In M. M. Webber (Ed.), *Explorations into urban structure* (pp. 19–41). Philadelphia: University of Pennsylvania Press.

———. (1996, September). Tenacious cities. Conference research notes: Spatial Technologies, Geographical Information and the City, Baltimore, MD. Retrieved October 3, 2006, from http://www.ncgia.ucsb.edu/conf/BATIMORE/authors/webber/paper.htm.

·2·

MUNICIPAL WI-FI COMES TO TOWN

Harvey Jassem

American cities are going Wi-Fi. They're moving to bring the information highway to their residents, businesses, and visitors. Good stuff, it seems, but not without important and sometimes controversial issues being raised along the way.

The information highway, in this case, refers to broadband wireless access to the Internet. That could result in all of the information, entertainment, and resources of the Internet being more readily available in cities. And as nearly any sort of information can be digitized and shared, it means that people and organizations could send and receive an enormous range of things with fewer technological barriers than ever. Whose role is it to foster such innovation? The American economy is a mix of private and public enterprise. There are those who suggest that the information infrastructure should be built entirely by private entrepreneurs; others suggest that municipalities should take on that role. This chapter examines the development of municipal government-enabled Wi-Fi systems in the United States.

What Is Wi-Fi?

Wireless fidelity, or "Wi-Fi," refers to an unlicensed interoperable local area network that permits simultaneous broadband access. The specifics of the technology are not important for purposes of this discussion. They will change over time. Newer wireless technologies, including what is being called Wi-Max (worldwide interoperability for microwave access), are likely to replace the current Wi-Fi standards, and such changes will bring with them issues relating to standardization and forward/backward compatibility, but for the purposes of this chapter the term *Wi-Fi* is used to describe both the present technology and similar yet-to-be-developed technology supporting wireless broadband access.

Wi-Fi and broadband are channels designed to carry a great deal of digital information. They may be used for closed networks—such as the one a residence or business might have allowing its computers to connect to printers and to each other. Or they may be open networks—designed to connect to other users through such facilities as the Internet. Finally, there are hybrid systems that are closed to outsiders but permit users to access outsiders. A password-protected system that is connected to the Internet would be a good example of that. For most uses, while broadband is the channel of communication, computers are the encoding and decoding devices that process the information sent through that channel. Occasionally—and this is likely to become more prevalent—other devices such as personal digital assistants (PDAs) and cell phones will be able to use Wi-Fi networks as well.

Where Is Wi-Fi? Where Is Broadband?

Broadband, which is most widely available via hard-wire connections (generally from cable television or telephone companies), is growing in availability fairly quickly. According to the Federal Communications Commission (FCC), broadband penetration has been growing at a rate of over 30 percent a year since 2000. There are approximately thirteen subscribers per one hundred U.S. inhabitants, and while that reflects significant growth, the United States is only twelfth on the list of per-capita broadband penetration of the Organisation for Economic Co-operation and Development (OECD) nations (OECD, 2004). Broadband is typically more available and used more in urban areas than in rural areas (FCC, 2006, p. 4). It is also used more often in higher-income neighborhoods than in lower-income neighborhoods. Because it is costly to distribute the broadband "wires," telephone and cable companies are understandably reluctant to bring it to sparsely populated or low-income areas where there may be too few customers to make such an investment profitable. And the costs associated with subscribing to broadband, which generally range from $200 to $600 per year for a single household connection, tend to preclude many lower-income households from subscribing. The expense of the necessary computer adds to the problem.

Broadband is available outside of the home in a growing number of locations. Libraries, schools, and businesses often have broadband available for their employees and clients. So-called Internet cafés have served as something of an Internet broadband equivalent to the public coin telephone. Customers can walk into such a "café" and log on to a broadband Internet-connected computer and pay by the minute for such access. All of these public access points require users to physically go to them in order to access the broadband connection, and many restrict access to sites they deem inappropriate for one reason or another. But as the explosive growth in cell phones has demonstrated, people would rather not have to go someplace in order to find a telecommunication station. People want full choice and not edited versions of the World Wide Web. And, increasingly, people want their Internet wherever they happen to be.

The demand for widespread and ubiquitous Internet access is strong. Businesses want it to be in easy reach of their suppliers and customers. Governments want it in order to do the business of governing and to be available to constituents. Individuals want it in order to have access to the huge range of information, entertainment, and other individuals available. There is no question that there is high demand for broadband Internet access. The question is, who will address that demand and by what means?

As previously noted, telephone and cable companies have been the primary sources of broadband. In the vast majority of cases, they do so as for-profit entities. In addition, there are other nascent broadband suppliers, including electric or power companies that are beginning to carry broadband over power lines, and satellite communication companies. It is reasonable to note that the private enterprise system is working and that private firms are making broadband available to address the need—at least the need of those individuals who live in served areas and who can afford the service.

Yet in a rapidly growing number of municipalities around the United States, local governments are getting involved in the development of wireless broadband access, either as owners and operators of the system or as partners with private firms. This is what is known as "municipal wireless," "muni wireless," or "muni Wi-Fi."

Why Muni Wi-Fi?

Nobody doubts that cities are already complex entities. Why would they look to get into the business of providing Wi-Fi? The suggestion that residents and businesses want it is insufficient. People want a lot of goods and services that government does not supply. There are several rationales for supporting municipal provision of wireless.

Wi-Fi is an electronic equivalent of a roadway, and there's a long history of municipalities providing the roadways that are necessary for commerce, resident convenience, and quality of life. It is a resource that will be used by the masses for trade, work, school, and other endeavors that represent the common good of the community. One cannot grow a community without roads, and this electronic equivalent of the roadway system could be a public good that is essential for civic growth. In this view, it is in the interests of the city for everyone to have access to such a resource. Much of the value of a telecommunication system is in its ubiquity. Value-added providers, those who will come in and offer services that rely on the communication channel, are more likely to invest in such offerings if "everyone" has access to the channel.

Municipal governments themselves could operate more efficiently if they had ubiquitous Wi-Fi available. For example, presently many emergency service personnel rely on cell phones and expensive cellular broadband data options that could be largely eliminated if ubiquitous Wi-Fi were available. Wi-Fi could make it easier for traffic departments to control traffic signals, for building inspectors to review city records while out in the field, for police officers to access records and file reports from the field, for parking meters to "call in" when they're malfunctioning (or accept electronic payments from cars), and so on. Municipal school systems could save on networking expenses and make networking more available at schools with Wi-Fi. Furthermore, Wi-Fi could help municipalities offer more government services and resources to residents. Meetings and files might more easily be shared with residents via Wi-Fi. This greater efficiency not only can result in more open government, it might also make it possible for more residents to serve themselves rather than using the time of city employees.

As a major source of information, governments need to make that information available to residents and businesses as quickly and widely as possible. Increasingly that means via the Internet. Equitable access to the Internet may require government intervention.

Governments legitimately look to create and support conditions encouraging economic development. They build or offer special zoning or tax incentives for industrial parks and other such economic development engines. Low-cost Wi-Fi is a business incentive that a growing number of firms want and municipalities can try to provide. This electronic highway system becomes increasingly important to business as we move further and further into the information economy. Moreover, important infrastructure might best be built by public entities that will support them rather than relying on private enterprises that might abandon them when they see a better place to invest their resources (Barbano, 2006).

Governments have a long history of providing or being involved in the development and provision of natural monopolies, public utilities, and telecommunication. There are government-operated power companies, and water and sewer operations. Natural gas firms typically get government easements and are regulated both in terms of price and service as a natural monopoly. Cable television firms are generally subject to local franchise regulation and are sometimes municipally owned and operated. When such services are deemed to be important utility-like resources, it is in the city's interest to ensure reasonable access, price, and acceptable/desired levels of quality of service. A case could be made that the electronic highway, as represented by broadband, is a natural extension of these traditional municipal concerns.

Governments play a key role in public safety, welfare, and education. They provide emergency, safety, and health services. Wi-Fi can support all of these. It can help to solve some of the problems encountered when different emergency responders use incompatible radio communication systems. It can be used by emergency responders (or by residents) to quickly transmit important medical information from homes to hospitals. It can serve as a redundant telephone system. It can tie communities together in an empowering fashion.

To the extent that muni Wi-Fi helps to bring the disenfranchised into the ranks of the enfranchised, it may work to reduce poverty, crime, and social isolation and alienation. These again are social goods that some in government wish to address.

The Muni Wi-Fi Playing Field

A growing number of municipalities are getting into Wi-Fi. Many of these are small communities that already operate municipal electric utilities, some of which are offering telecommunication services. Larger cities are getting into the business as well. In 2005 there were twice as many municipalities getting into Wi-Fi as there had been a year earlier (Lehr, Sirbu, & Gillett, 2006). Though the reasons for municipal involvement with the development of local Wi-Fi may be compelling, it is not without risks or powerful countervailing forces.

Building a citywide Wi-Fi system is expensive, and critics of muni Wi-Fi note that the demands on tax revenues are already substantial. Building a municipal Wi-Fi system, they say, will have to be at the expense of some other valued municipal services. Should funds be diverted from the school systems to support the muni Wi-Fi? From the police or fire departments? Faced with such difficult decisions, some in government resist calls for municipal involvement. Others note that widespread urban Wi-Fi systems are still something of an unproven novelty and question whether cities should be on the cutting edge of the risk takers ("Wi-Fi Pie," 2005).

Similarly, some observers simply don't believe that government should get involved or provide products and services that the private sector can provide. They tend to be sympathetic with the private market forces that would have government step away from all but the most essential services. For example, Robert Sahr, chair of the South Dakota Public Utilities Commission, testified to the U.S. Senate Commerce Committee that "before pursuing a municipal-owned or sponsored network, we should look first to private solutions. Our nation's telecommunications providers have made substantial investments. . . . We must be particularly mindful of municipal action that displaces or discourages private investment. . . . Without some type of market failure, municipal entry . . . is highly suspect" (Senate Commerce, 2006). The New Millennium Research Council (2005), a research group backed by the telecommunications industry, charged that muni Wi-Fi would have "a detrimental effect on city budgets and on competition in the telecommunications industry" (p. viii).

Still others in government argue about the correct site for government intervention in Wi-Fi. Should cities and towns have the authority to build and/or regulate Wi-Fi and broadband? Should that power go only to the states? Should the FCC or Congress decide these issues? Predictably, each level of government has advocates for making that level of government the proper site for such oversight and authority. In the complex world of telecommunication regulation, the locus of government authority is a reasonable question. Existing regulatory and government oversight of other forms of telecommunication does not adhere to a consistent overarching philosophy or structure. Wireless broadcast television is regulated by the FCC at the federal level. Cable television is regulated by a mix of federal, state, and local entities. The same functions provided by "cable" and by "telephone" companies fall under different regulatory models. The telecommunications field is characterized by regulatory incoherence. The case of Wi-Fi does not scream out for one clear regulatory principle.

The discussion, in its broad terms, is quite old. The municipal electrification movement provides some interesting parallels. In the late nineteenth century, as electricity was coming to American towns and cities—not quickly enough for some—with the promise of great change, many municipalities formed entities responsible for bringing electricity to their communities. Technological advances were making the farming industry less labor intensive, resulting in a migration of workers from rural areas to cities. Urban industrialization provided much of that work. Cities and towns found themselves rushing to provide the infrastructure necessary for their survival and health. So too were private entrepreneurs. In 1899, economist Edward Bemis wrote: "Three great questions now confront us: shall we have public regulation [of utilities], or public ownership? If the former, what shall be the nature of the regulation? If the latter, what are the dangers to be avoided?" (p. v).

Early proponents of municipal control of the infrastructure (roadways, bridges, public transportation, water, sanitary facilities, and electricity) sometimes noted the corporate corruption being spotlighted in the popular press of the day as one of the reasons for moving these essential services out of the hands of private entrepreneurs and into the hands of the government. It was a heated political issue with a number of different models emerging. One of the more interesting and relevant outcomes was the development of state regulation of electricity as opposed to municipal operation of the electric utility. State regulation was promoted by state leaders who did not want to give up power (excuse the pun) to municipalities, and by entrepreneurs who wanted to thwart muni electric providers that might compete with their own efforts to get into that business (Schap, 1986). Muni electricity faced many of the same criticisms facing muni Wi-Fi. Critic M. J. Francisco wrote in 1893: "Strange as it may seem, the original command, 'Let there be light!' still needs to be enforced; this science fraught with so much good for mankind, is now menaced by men, who, posing as economists, desire to use this power for political purposes, as well as for their own aggrandizement, regardless of the interest of the masses or the public welfare. The plain English of the movement is, municipal control of electric lighting plants" (p. 9). Francisco went on to argue that municipally owned electric power systems would be far more expensive to operate than its proponents suggested and that advocates of government ownership were disingenuous power seekers. Through the early twentieth century, the debate raged on in America. Ballot initiatives and legal battles were fought over the concept of muni electric versus private ownership of electric power plants. In 1930, Frederick L. Bird and Frances M. Ryan looked at the young history of these battles and noted that proposals for muni electric had been “fervently opposed as ‘rat holes for the tax-payers’ money’ and as the dread harbingers of socialism” (p. xv). Yet their study of muni electric systems in California at that time suggested that such systems were efficient, especially in the larger cities, they generated revenue for the cities, and their low rates to consumers served to keep down the rates of private electric companies.

Predictably, a 1928 book on the topic of muni electric that was published by the National Electric Light Association, representing private power companies, did not support such favorable assessments. It noted that muni electric “is government in competition with its own citizens” (p. 3). The book turns to earlier history in American infrastructure development for support for its contention that government should stay out of the way and let private industry do the building. More specifically, it turned to the history of railroad and canal construction—particularly interesting because these were characterized as means of “internal communication”—and suggested that government-built canals and railroads, while meant to support economic development, most often wound up as embarrassing boondoggles. It turns for support to President Coolidge’s 1928 remarks to the Daughters of the American Revolution that “our theory of society rests on a higher level than communism. We want the people to be the owners of their property in their own right” (p. 51). It also turns for support to Herbert Hoover’s 1924 response to Senator La Follette’s call for public ownership of utilities, noting that the “Republican Party stands for private ownership. . . . Either we are to remain on the road of individual initiative, enterprise and opportunity . . . , or we are to turn down the road which leads through nationalization of utilities to the ultimate absorption into government of all industry and labor” (p. 55). There was no shortage of strong rhetoric opposed to government ownership of electric power plants or any other utility.

Municipal provision of electricity did not follow a single model. Some municipalities provided the power generation, transmission, and distribution of electricity. Others, especially as the private industry became more robust, bought the electricity generated by third parties and distributed it locally. And not all municipal electric operations operated long term. In the 1920s and ’30s, some municipalities that had gone into the business of muni electric turned the operations over to private enterprise in order to take advantage of newer economies of scale or to get out from under the constraints of being able to operate only within their political boundaries. From 1923 to 1927, more than one thousand muni electric operations changed to private ownership in the United States (Schap, 1986). Interestingly, this trend toward privatization was stemmed in the 1930s as the United States experienced an economic depression, governments were unable to effectively curb private firm rate increases, federal government programs provided funds for municipal projects, and federal law required that excess electricity produced by new flood control dams be sold to municipalities at rates lower than it would cost them to generate their own electricity. As the growth in federal relief programs slowed, as muni electric plants aged and became relatively inefficient, as the economy grew anew, muni electric was largely replaced by private electric.

The arguments around the "proper" role for government as a utility provider today persist, though with a bit more subtlety and complexity. Politics and government have always been about the negotiation of conflicting values and resources. Government decision makers are often dealing with clashing views of different constituencies. Today's stakeholders are complex and have a variety of interests. There are a number of constituencies that do not favor muni Wi-Fi or favor it only in ways that will benefit themselves. They include private entrepreneurs who are now or might plan to be offering their own broadband access services. This could include cable and telephone companies (those that offer competing broadband service and phone service), cell phone companies (especially those offering 3G wireless broadband service, but even those that simply fear a loss of traditional cell phone business if people replace some of their calling with voice over Internet protocol [VoIP] via Wi-Fi), satellite telecommunication companies, fixed wireless multichannel multipoint distribution service (MMDS) providers, electric power firms (those that see the potential to offer broadband via power lines), and the like.

In general, those opposing government intervention into the market characterize such an incursion as unfair competition. Douglas Boone (2006), a telecommunications CEO representing the U.S. Telecom Association to that same Senate Commerce Committee, said, "government owned networks are not akin to other public utilities. In fact, government networks are more akin to City Hall opening a chain of grocery stores or gas stations. They typically require heavy taxpayer subsidization . . . benefit from tax advantages and regulatory exemptions that do not apply to private firms . . . are not subject to the pressures and stresses of the marketplace [and] often neglect innovation" (pp.

1–2). He asserted that his company lost nearly 50 percent of its subscribers in a town in which the local government built a competing network and threatened to raise taxes if residents didn't support the municipal network. Note that Boone's claim that muni Wi-Fi systems typically require "heavy taxpayer subsidization" is not necessarily true. Muni Wi-Fi systems are just starting to come on board in larger communities, and the financial relationships they have with the cities increasingly appear to promise something else.

While firms that might compete directly with muni Wi-Fi systems could be expected to oppose such operations, other firms that might serve or take advantage of them logically support them. Chris Caine (2004), vice president of government programs at IBM, has called on governments to invest in and develop advanced and open telecommunications infrastructures "the same way governments invested in infrastructures for the industrial age" (pp. 237–242). He suggests that governments could help grow a robust information economy and enhance national security by building such systems themselves, by creating policies that encourage private investment, and/or by creating public-private partnerships. The market for muni Wi-Fi is likely to be huge. Even one-time opponents to muni Wi-Fi are starting to try to get a piece of what is predicted to be a $1.2 billion market by 2010 (Sharma, 2006, p. B1).

In marketplace economics, it is understandable that firms would be happier without competition of any sort. That is not the same as wishing for an absence of government intervention. In the area of utilities and telecommunications, firms often seek government intervention to ensure their monopoly position, reasoning that the large capital investment required to build a citywide system might be justified only if there is some guarantee of protection from competition. Recall the dire predictions of incumbent telephone companies when their government-protected local monopolies were opened up to competition. Many of the parties opposed to government operation of Wi-Fi would be similarly opposed if the newcomer were named Acme instead of City Hall. Still, there are legitimate concerns about a level playing field when one of the players is the entity that makes many of the rules and owns much of the infrastructure upon which the networks are built.

Regardless of whether or not Boone's firm lost 50 percent of its subscribers, or whether government competition was the reason, or even whether that might have been a good or bad thing from a larger perspective, it is clear that there are interested parties who fear government intervention into the broadband market, and they are willing to lobby to protect their interests. A 2006 *Telecommunications Policy* article declares that U.S. "broadband diffusion is currently at the regulation crossroad between the market, government and consumer lobbyists" (Papacharissi & Zaks, 2006, p. 64).

As fast as municipalities issue requests for proposals from potential partners interested in building/operating muni Wi-Fi systems, others are trying to shape the future of muni Wi-Fi development in statehouses, courts, federal agencies, and Congress. A number of states have enacted laws prohibiting municipal ownership of Wi-Fi or telecommunication networks, Congress has considered bills that would both prohibit and permit such municipal ownership, and the U.S. Supreme Court has ruled on states' authority to limit such municipal ownership (Community Broadband, 2005; Nixon, 2004; Peirce, 2005; Preserving Innovation, 2005; Stone, 2005).

A Sampling of Cities Entering the Market

Philadelphia: Subscriber-Based Business Model, Public-Private Partnership

Philadelphia is the first large U.S. city to enter into an agreement to build a muni Wi-Fi system. The city established a nonprofit corporation, Wireless Philadelphia, that will oversee the construction and operation of the system that will be built and operated by Earthlink, a for-profit company that previously focused on providing dial-up Internet service. Now that dial-up service is being abandoned in favor of broadband, and because cable and DSL broadband providers are permitted to discriminate against other Internet service providers (ISPs) wanting to do business through their facilities, Earthlink is seeking new business models, and the Philadelphia story represents one of them. The City of Philadelphia will not invest financially in this Wi-Fi build. Earthlink will build the system (initial estimates put the cost at around $10 million) and will pay the city an up-front fee as well as annual fees for the use of its streetlights (on which Wi-Fi equipment will be mounted). Earthlink must also share some of its revenues with Wireless Philadelphia (which, in turn, must pay the local electric company for the electricity Earthlink's Wi-Fi equipment uses). The contract also requires that Earthlink offer up to 25,000 subsidized low-cost ($9.95/month) accounts. Earthlink must permit other ISPs to have access to its Wi-Fi customers, but Earthlink may set the wholesale price it charges those

ISPs. Wireless Philadelphia has the right to buy the service at prenegotiated wholesale rates and resell it retail to end users. This could help keep Earthlink's retail prices in check. In addition, the contract has benchmarks for acceptable levels of reliable service and Wi-Fi penetration throughout the city. In what Becca Vargo Daggett (2006) calls "a clear bow to cable incumbent Comcast, and perhaps Verizon, . . . Earthlink may not offer cable or online video service."

The City of Philadelphia, then, will get citywide Wi-Fi that is available for reasonable—and in some cases subsidized or free—rates. The city is not investing financial resources into the system and will receive modest income from it.

While Philadelphia was working to develop its muni Wi-Fi project, the Pennsylvania legislature passed a law prohibiting other municipalities in the state from following suit.

San Francisco: Subscriber- and Advertiser-Based Business Model, Public-Private Partnership

After Philadelphia constructed its deal, San Francisco issued a request for proposals (RFPs) and sought providers who might agree to provide Wi-Fi service for free to the city's residents. The notion that the muni Wi-Fi be available to all for free was pushed by San Francisco's mayor, who positioned Wi-Fi access as a basic human right and led the initiative. As has been the case with many of the municipal RFPs, there were many submissions. Ultimately the city accepted a proposal that came jointly from Earthlink and Google. Essentially, it provides for a system whereby the companies will build the system (at an expected cost of approximately $15 million) and offer users a choice of tiers. Free access will indeed be offered, supported by advertising. This level of access will be slower than most broadband but faster than dial-up. Faster Wi-Fi service will be available for a fee expected to be about $20 per month.

Boston: Independent Network, Private Internet Service Providers

Boston may be taking an altogether different approach. As is often the case, the mayor formed and charged an independent Wireless Task Force to come up with a recommendation for bringing Wi-Fi to the city. The task force included members from the city government, business (though not ISPs), academia, the community, and professional consultants. Their plan, supported by Mayor Menino, would establish a nonprofit entity that would build the Wi-Fi network system, funded by foundations and businesses, with goals of

- Promoting economic development and stimulating innovation
- Ameliorating the digital divide
- Improving the quality and efficiency of City services. (City of Boston, 2006, p. 6)

The cost to build the system is forecasted to be between $16 and $20 million. This network and the organization responsible for building it, unlike the others discussed here, would not offer Internet access. Rather, it would rely on other firms to use the network and offer their services as ISPs. The task force is counting on the idea that a bevy of competitors would line up to offer such services and that, as a result of that competition, prices and service packages would vary broadly. Perhaps some would offer free advertiser-supported systems, others might offer pay-per-use options, and others might offer different speeds for different prices. The Boston task force speculated that paid access might cost subscribers $15 per month. The system would be built on city-owned buildings and light poles, thereby keeping costs of construction low. "What this will do is give us citywide service at a reasonable cost," Menino said (Weisman, 2006, p. A1). "This is a unique approach. We're not turning it over to someone else. We'll be able to control our destiny. One outside corporation shouldn't have a monopoly over this technology" (p. A1).

The recommendation notes that while the system should start out as a Wi-Fi system, it might evolve to use other standards. It also notes that while ubiquitous coverage is essential, Wi-Fi is not very good at penetrating buildings and structures. Hence, it suggests a goal of getting the signals to the outside walls of all the buildings in Boston, and leaving it up to end users to amplify the signals inside their interior space.

This plan is still just a plan. It is also unique and potentially groundbreaking in the United States. As of this writing, no organization has been established to build the network. No funds have been raised to build the network. City leaders express optimism that all of this can happen in a matter of months. It is not yet clear what actions the existing broadband suppliers in Boston might take.

Because all of these systems are newly contracted (or presently in negotiation), it is possible that the terms of the agreements may change. If the firms building these systems begin to have financial problems, see that their financial models aren't panning out, or find themselves in a rapidly changing marketplace, they might pressure cities to renegotiate these deals. That is what many cable television firms did after winning franchises to provide cable television service. It should not come as a shock if something similar develops in the area of muni Wi-Fi.

The Future

Muni Wi-Fi is in its infancy. The models of development will continue to change. Urban muni Wi-Fi started with subscriber-funded systems. The move toward advertiser-supported and municipality-supported systems reflects a willingness to consider other approaches. Additional models will emerge as well, complementing the changes in the industry, the technology, and consumers' and governments' expectations. That America Online switched from a subscriber- to an advertiser-supported service in late 2006 points to this change. Consumers have a growing number of entry ramps onto the information highway. Consumers are spending a growing percentage of their disposable time, attention, and money on the Internet. And the same is true for pretty much the entire range of organizations: businesses large and small, nonprofits, educational resources, and government entities. The "coalitions" of forces supporting muni Wi-Fi are shifting as well. The visions, the expectations, the power to affect change will continue to morph. And as these new business models are tried and tested, participants will see costs and benefits that are not yet apparent. The Earthlinks and Googles of the world may find that these business models don't work or that others provide greater promise. Advertisers, politicians, and end users will all gauge the success of these systems against their own goals and needs. New communication channels and uses will present themselves and pose new challenges. The promises made in these early experiments will be judged against their performance.

During the coming years, observers have the opportunity to study the impact of muni Wi-Fi and compare the development of competing models. They might ask:

- Is ubiquitous free Wi-Fi hastening the end of the digital and knowledge and socioeconomic class divides?
- Will the limited amounts of subsidized Wi-Fi accounts be sufficient to satisfy the need?
- Will certain systems be more subject to abuse and privacy violations and technological problems than others?

- Which models are able to sustain themselves financially?
- Which models best serve public needs?
- In what ways do political and economic alliances affect the development of muni Wi-Fi?
- What do consumers want, what are they willing and able to pay for it, and what do they need to take advantage of these resources?
- Which models seem the most valuable for which users/uses?
- What are the best predictors for success?
- What impacts will muni Wi-Fi successes and failures have on future development?

Many additional questions remain to be addressed. One thing seems certain: the genie is out of the bottle. Wireless broadband will become increasingly available, probably through a multitude of providers. And that suggests one additional area of inquiry: What is the impact on the economy and life of the cities in which there is widespread Wi-Fi?

There is an operating assumption in all of this that the spread of the information highway is inherently a good thing. And maybe it is. But it is also something that will change life as we know it, much as transportation highways and widespread electricity changed things. The Interstate Highway System certainly brought places and people closer together in many ways. But it also divided and in some cases killed neighborhoods and towns. That is the way with innovation. There are unintended and often unanticipated consequences. Will muni Wi-Fi bring riches to all of the people, or will it merely distract us from more productive ways of living in our communities?

References

Barbano, R. (2006). Commentary: Are we building information highways or railroads? Retrieved August 7, 2006, from http://muniwireless.com/municipal/1273.

Bemis, E. W. (1899). *Municipal monopolies: A collection of papers by American economists and specialists*. New York: Thomas Y. Crowell.

Bird, F. L., & Ryan, F. M. (1930). *Public ownership on trial: A study of municipal light and power in California*. New York: New Republic.

Boone, D. A. (2006, February 14). Senate Commerce, Science and Transportation Committee. Communications issues. Retrieved from http://commerce.senate.gov/pdf/boone-021406.pdf.

Caine, C. (2004). Government in an era of rapid innovation. [Speech delivered at the World Knowledge Forum, Seoul, South Korea, October 16, 2003]. *Vital Speeches of the Day*, 70(8), 237–242.

City of Boston. (2006). Wireless in Boston: Wireless task force report, broadband for Boston. Retrieved August 16, 2006, from http://www.cityofboston.gov/wireless/Boston%20Wireless%20Task%20Force%20Report%20-%20Final.pdf.

Community Broadband Act of 2005. (2005). S. 1294, 109th Cong., 1st. Sess.

Federal Communications Commission (FCC). (2006, July). High-speed services for Internet access: Status as of December 31, 2005. Retrieved August 28, 2006, from http://www.fcc.gov/wcb/iatd/comp.html.

Francisco, M. J. (1893). *Municipal ownership: Its fallacy. With legal and editorial opinions, tables and costs of lights as furnished by private companies and municipal plants*. Rutland, VT: Carruthers & Thomas.

Lehr, W., Sirbu, M., & Gillett, S. (2006). Wireless is changing the policy calculus for municipal broadband. *Government Information Quarterly*, 23(3/4), pp. 435–453.

National Electric Light Association. (1928). *Government (political) ownership and operation and the electric light and power industry*. National Electric Light Association.

New Millennium Research Council. (2005, February). *Not in the public interest: The myth of municipal Wi-Fi networks*. Retrieved March 19, 2007, from http://newmillenniumresearch.org/archive/wifireport2305.pdf.

Nixon, *Attorney General of Missouri v. Missouri municipal league et al.* (2004, March 24). Retrieved October 22, 2006, from http://a257.g.akamaitech.net/7/257/2422/24mar20041200/www.supremecourtus.gov/opinions/03pdf/02–1238.pdf.

Organisation for Economic Co-operation and Development (OECD). (2004). OECD broadband statistics, December 2004. Retrieved August 8, 2006, from http://www.oecd.org/document/60/0,2340,en_2825_495656_2496764_1_1_1_1,00.html.

Papacharissi, Z., & Zaks, A. (2006). Is broadband the future? An analysis of broadband technology potential and diffusion. *Telecommunications Policy*, 30(1), 64–75.

Peirce, N. (2005). City-sponsored "wifi"—saved from the telecoms? Retrieved August 21, 2006, from http://www.napawash.org/resources/peirce/peirce_08_21_05.html.

Preserving Innovation in Telecom Act of 2005. (2005). H.R. 2726, 109th Cong., 1st Sess.

Schap, D. (1986). *Municipal ownership in the electric utility industry: A centennial view*. New York: Praeger.

Senate Commerce, Science and Transportation Committee. (2006, February 14). Communications issues.

Sharma, A. (2006, March 20). Companies that fought cities on wi-fi, now rush to join in. *Wall Street Journal*, p. B1.

Stone, M. (2005). Wireless broadband: The foundation for digital cities: A cookbook for communities. Alpharetta, GA: Civitium. Retrieved October 10, 2006, from http://www.mediasf.org/index.php?module=uploads&func=download&fileId=52.

Vargo Daggett, B. (2006). Wireless Philadelphia–Earthlink contract: An analysis. Retrieved April 26, 2006, from http://muniwireless.com/municipal/bids/1151.

Weisman, R. (2006, July 31). Hub sets citywide wifi plan. *Boston Globe*, p. A1.

Wi-Fi pie in the sky. (2005, March 19). *Economist*, 374, 78.

·3·

MOBILITY IN MEDIAPOLIS

Will Cities Be Displaced, Replaced, or Disappear?

Gene Burd

New mobile communication technologies are the latest preoccupation of the perennial predictors of the demise of the city as a geographical place. The possible death of the physical city has been a frequent fear, from the fall of ancient city walls to the gated cities of today; in between, there have been the fateful forecasts for the end of the city since the Industrial Revolution, the separation of work and residence, the rise of the auto and suburbs, and the predictions of the unsustainable city in a deteriorating natural environment.

Although "our entire history is connected to space and place, geometry and geography" (Negroponte, 1996, p. 238), dread continues that cities will disappear because of social, economic, political, religious, military, and natural forces, and changes in communications technologies—both personal and mediated. Adaptations have been made in cities by print and electronic mass media with fixed points of origin, which have changed urban social geography. However, with ubiquitous and portable 24/7 mobile and wireless communications, such as cell phones, pagers, iPods, BlackBerry devices, and laptops, that geography has been placed in motion, with a new challenge to the city whose physical site has anchored its traditional locale and whose stationary communications have largely defined it as a city.

In today's technological flux and uncertainty, "anyone who promises to foretell the detailed 'trajectory' of mobile telephony should probably be disregarded out of hand" (Rule, 2002, p. 254). Therefore, it would seem an appropriate time to renew discussion of the nature and future of the city as a geographical, digital, virtual place amid its continuing decentralization by multiple centers, multiple cultures, and multiple media. This is all the more important because of the "disquieting void" that research "experts have largely ignored" with "slight academic interest" in the communication "taking place right in their own ears" on mobile phones, whose universal use exceeds that of television or the Internet, especially in less modern oral cultures (Katz & Aakhus, 2002, p. 3). Even Erving Goffman, "arguably the most astute observer of the routine and the mundane, seldom talked about the phone" (pp. 9–10). Over one thirty-year period of research, there were fewer than five articles on CB radio, but an average of five articles published daily on computer-mediated communication (Katz & Aakhus, 2002).

While the study of the social impact of the telephone has been assessed (de Sola Pool, 1977), the emerging sociology of its mobility offers yet another useful avenue to view the city as a place where public and private communication overlap, where physical proximity has been an evolutionary precondition for social interaction, where a stable place has been necessary for complex communication, and where cell phones lessen the degree of spatial anchors for social relations and decrease the positive impact of spatial proximity in social interaction (Geser, 2004). The relation of close proximity and fixed stationary communications to sites may become even more crucial as media content becomes more virtual and sites more mobile, because the city as a place already has been considerably displaced by mediated images and representations, and mobility may move the stationary city further from the experience of time and place. The mobile phone has married the auto, and provides an apt communications study site where perhaps dramatic news stories on drivers using cell phones may finally convince media scholars that automobility—transportation and communication—are indeed inseparable.

There are early clues that new mobile communications, linked to the computer and Internet, may not necessarily erase the city as a place, despite the frantic fears of nostalgic technology critics and the hyped hopes and self-interests of the technology apologists. The city as a place may actually become more important, although its form will likely be modified, while its communication functions remain and in the process give rise to new definitions of the city. New mobile media could make the city a victor rather than a victim, as the wireless world of a mediated "city of bits" and bytes (Mitchell, 1995) mixes and melds the virtual and real, forces old media to merge and readjust, and even leads to the rediscovery and recovery of the interpersonal communication considered inherent in the historical reason for cities to exist.

The changing city does appear to be moving more toward the virtual mediated metaphorical and mechanical mental images that continue to be projected from its geographical spaces. During such movement, the potential for interpersonal communication is often minimized or suspended and the created spatial distance is often incompatible with communication (Geser, 2004). In city sites of postmodern urban space, "it is difficult to find a fixity or stability in identity, as individuals inhabit many differing identity positions," which are fleeting and private and with temporal and spatial fluidity (Drzewiecka & Nakayama, 1998, p. 29). Mobile communication is relocating place to "nonplaces" unrelated to the messages and messengers who are involved, and content is being determined by participants and not by the setting (Augé, 1995). Meanwhile, the virtual city is being reconfigured by obsessive and excessive predictions of the end of history and geography by technological determinists impressed by the myth, mystique, and magic of cyberspace (Mosco, 2004). In addition, the limited and neglected research and theory on telephones and cities may be more misplaced than the realities of any displaced or replaced city, whose communication function persists while its form continues to change.

The city has long been the prime arena for historic clashes over space, place, and communication. From writing to wheels, from caves to computers, from Gutenberg to Google, from walled to gated cities, changes in communication and transportation have not replaced cities, which have survived as a geographical and spatial entity and as a mediated virtual and symbolic memory even when their geographic sites have disappeared. When defensive physical city walls were built and then declined after the era of Greek, medieval, and Renaissance cities, some suggested that the "true" city ceased to exist when citizens ceased to guard its walls (Martindale, 1958). "The City Is Dead—Long Live the City," observed one urban scholar, who suggested that "as the city changes, so do the terms used to describe it" (Long, 1972, pp. 24–26). She noted that the city's "freeing of interaction from its geographical" face-to-face base had created a new, odd, different, and ignored secondary, tertiary, superficial (virtual) interaction within the physical and social distance resulting when the city's communications grid reached beyond its transportation web (p. 4). That new city without walls had expanded because of commerce and capitalism, which "turned local citizens into consumers" in a geographical market and "real estate site" with symbolic walls that did not coincide with or command local city loyalty for a shared common enterprise (pp. 25, 54). As mobility has dissolved time and place, architecture and history in the city have shifted the control of time and place from the fixed temple and its records to individual clocks, calendars, and now cell phones. Thereby, "life in the real-time city" of "mobile telephones and urban metabolism" (Townsend, 2000) becomes "the city in your pocket" (Kopomaa, 2000). Preservationists and city planners have long pondered and wondered, "What time is this place?" If time were not embodied in the physical world, and if the virtual and ephemeral replace space and time, then the "framework within which we order our experience" of places precedes any memory in cities as negative feedback negates it (Lynch, 1972, pp. 241–242).

Add to this changing pattern of places the shifting and porous boundaries and margins of nation-states with countries and cities as potential casualties on obsolete cultural maps redrawn to keep pace with immigration and quick transportation. Dislocation and disasters, combining human and natural, have also revived fears of urban decline and demise in regard to the stability, utility, and futility of cities, and even the planet, but cities still survive the plagues, famine, fires, and terrorists (Henricks, 2002).

The city as a place still matters, although its shape and domain have changed. Emerging urban and environmental regions are persuading planners and designers to project new structural images of space—both built and natural—in order to create a new and changing sense of urban place (Burd, 2002). Advocates of a new "metro-politics" contend that while "technology appears to be conquering space" for most affluent and mobile urbanites, "place-based inequities" remain for the poor who are "left behind" in "many places" where their residence affects their economic segregation in "places becoming economically isolated" (Dreier, Mollenkopf, & Swanstrom, 2001, p. 1). For them, cities as "places of intense personal interaction" are as important as ever, and the idea that the city is obsolete is "nonsense." They lack access to "computers, the Internet, e-mail, cable, satellite dishes, faxes, mobile phones, and other new media technologies which would enable them to access many of society's benefits" (p. 2).

New media have not necessarily erased the "where" and "here" of place: Alexander Graham Bell's first phone message in 1876 was "Mr. Watson—Come here, I want to see you" (Mitchell, 1995, p. 35). Place mattered to printer and publisher Brigham Young, who felt sure that "this is the place" for Salt Lake City. Poet Gertrude Stein failed to find Oakland, California, because "there is no there, there." In today's "vanishing city," "there is less and less there anywhere, anymore" because "there is everywhere" and "one place looks pretty much like any other" (Pascal, 1987, pp. 600, 602). Recall even the first words uttered by space traveler Neil Armstrong taking that first physical step on the dusty moon surface, when he uttered the name of a city: "Houston, Tranquility Base here."

While mass media from the moon or from Manhattan deliver planned and prepackaged content, new individualized, democratic, and mobile communication technologies encourage users to design their own cybernetic "city" places and to connect to their fragmented communities of interest, freed from the restrictions of time and space. The cell phone is more than mere talk in the public square, as it combines interpersonal and mass media (O'Keefe & Sulanowski, 1995), and being immediate, anywhere, anytime, it transcends time and space, and dissolves boundaries (Leung & Wei, 2000), although mediated virtual reality is not so recent. The Great Wall of China (the only built Earth artifact visible from outer space) and early Chinese city walls were virtual as well as geographic boundaries whereby "paintings of walls served the same purpose as walls themselves" (Steinhardt, 2000, p. 459). They divided private and public worlds and, in later urban civilizations, even the words for "wall" and "city" became interchangeable (Tracy, 2000).

Places and media may be displaced, but not necessarily erased or even replaced. Each new medium can alter a sense of the city as a place and disconnect people from public place. More frequently, a new medium changes or modifies a previous one, rather than eliminates or makes the city obsolete. Both obsolescence and fragmentation are interwoven into the present and future. Each new technology tends to alter, reshape, and, in some cases, obliterate previous media or make them obsolete (Gumpert & Drucker, 2006). The mobile phone, for example, is being modified by convergence with other forms like photo, text, the Web, and broadcast, with mixed blessings for cities. The diffusion and municipal adoption of broadband and wireless fidelity (Wi-Fi) "may take us another step toward the de-urbanization of America" and "further urban sprawl" of cities, but "WiFi may both be necessary for their survival and result in their functional borders being spread beyond their political ones" (Jassem, 2005). Convergence is crucial when new convergent media devices enter public places and cities wrestle with the impact of the new media. "Although interaction has been emancipated from place, public places still function as sites for face-to-face interaction," and the physical proximity of individuals matters less to communities than in the past because modern media have eliminated cohabitation of a territorial place as a prerequisite for community (Drucker, 2006; Drucker & Gumpert, 1991).

Displacement theories of places are dependent on how cities are defined as a mixture of "real and true" (interpersonal, face-to-face) or as virtual and artificial (mediated, imagined). While media can separate people from geographical public space, they also can bring people back to that interpersonal space. And because a city is a communication network, new communication technologies do not automatically replace or displace the city, but can actually enhance and preserve it as a communication entity. The city as a place persists, with newly shaped concepts and containers assigned the metaphor of cyberspace, which creates new borders, boundaries, and margins using a mixture of old and new language. Even the geographic mobility of communication does not eliminate the city as a form and forum (both real and virtual) for the function of communication. Some would argue that mobile media might even rescue and save the city by restoring interpersonal communication to a manageable, metamorphic, and selective scale (Matei & Ball-Rokeach, 2003; Wellman, 2001) and lessen the interactive overload. As for the virtual nature of media, there is evidence that the "city of the mind" has long been virtual and *media*ted long after Georg Simmel's 1903 observations in "The Metropolis and Mental Life."

City forms change, but their function remains as the virtual and physical are interwoven. Walls fall, but new ones are erected with new margins and boundaries, and public and private turf is redefined and renamed. New places and spaces are created as new technologies reconstitute, alter, and augment the city to create a "tertium quid city" (Gumpert & Drucker, 2006) with physical and electronic parts but with a related "third thing" different from its sum, thus preserving the old with the new, unlike the new recombinant city or the digital city. Cell phones have similarly been called a "third skin," an extension of the self, and iPods may now be to people-watching what sunglasses used to be. This shift from public to private sociability may provide the same sense of security and comfort received from others in preindustrial villages (Visser, 2005).

Evidence indicates that the Internet brings people together and extends community rather than isolates people, and mobile phones connect people through networked individualism rather than through fixed and grounded groups. Before the 1990s, places were connected by cars, planes, railroads, mass media, and fixed telephones, but now people can be connected via the mobile phone and Internet (Wellman, 2005). Mobile communication liberates people from places and shifts community ties from linking people-in-places to linking people wherever they are to create "truly personal communities" (Wellman, 2001, p. 240), and the virtual communities in cyberspace are much like those that people develop in their "real life" communities (Wellman & Gulia, 1996, p. 15).

In this sense, perhaps the traditional city does become a victor more than a victim. Renegotiation takes place in the debate as to what and where the city is, as the old geography is reshaped and fractured communities of interest and geography conflict and coalesce, as with the Wi-Fi interface between the physical and the media environment. It is a crisis of interaction and identity involving distance and time, space and place, site, size, density, and the urban "way of life." Mobile phone users are communicating even though they are not associated with their physical place. "Their awareness and behavior is totally in private cyberspace even though their bodies are in public space" (Wellman, 2001, p. 240).

The flesh of the body has historically been situated in the stone of the city (Sennett, 1974), and that has implications for the bionic posthuman extension in the postmodern city as talk and speech communication could possibly return to local primary groups and away from mass media. Architecture could become even more virtual, as mobile cars and phones become combined. If oral culture is restored, communication boundaries could move, shift, and change. As old walls fall, fixed public clocks and statuary would not be the only ways to experience time and place in the city, as people share communal knowledge and culture and not just information and data.

As mobile communications move from places of fixed spaces of origin, the language of urban forms and places remains somewhat tied to the geography of the central city. Despite decentralization to suburbs, the hinterland, regions, and world cities, people continue to retain memory and the psychological need for symbolic mental maps.

This computerized virtual city "architecture" uses "virtual-place metaphors" (for example, rooms, dens, cafés, highways, dungeons) to preserve an already held "sense of self and society, of time and space, of community and cosmos" in an immaterial "sociotechnical system" that has attributes of place (Adams, 1997, p. 167). Perception and use of place and space are affected by these media, but city planners and media theorists have failed to integrate the impact of media on face-to-face urban interaction and to "choreograph both public space and the newer forms of electronic space" (Drucker & Gumpert, 1991, p. 305).

The fears of the new placeless virtual geography and architecture continue to resurface with the reluctant admission that "it's a sprawl world after all," which is seen as artificial and not "genuine," as unhealthy, lacking face-to-face contact, and a "breeding ground" for loneliness, depression, incivility, isolation, and violence (Morris, 2005). The sprawling and placeless model of Silicon Valley, "our new city," which is "without form, without any center of any kind," is deplored as a "real Nowhere Land" and "no place in particular" with "no there, there," a place "where you never arrive," and the "perfect example of the End of Place" (Marshall, 2000, p. 65). Aside from new cities "out there," within the emerging "digital cities" themselves (defined historically as physical places), there are internal battles and negotiations over places and spheres—public and private, physical and virtual, in the shaping of urban Internet space (Aurigi, 2005, p. 16). The new media technologies have generated "calls for redefinitions of place-making" with new cyber, virtual, and digital names for cities, villas, and towns (p. 17). This is not the first time this has occurred, because virtual buildings and cities were early represented in paintings and sculpture, but now virtual worlds and environments have moved beyond mere representation (Aurigi, 2005). The virtual is becoming the real.

However, the realities of geography and land are essential for both old media and the new cyberculture. Huge, sprawling hunks of land in the outlying rural Pacific Northwest are sought by Google, Microsoft, and Yahoo to build large structural global data complexes (Markoff & Fackler, 2006). The huge "googleplex" in Silicon Valley is itself a geographical spot. So are transmission towers and modified public buildings to accommodate electronic technologies. In urban areas, global positioning system (GPS) and global navigation satellite system (GLONASS) signals are often blocked by the geographical realities of urban canyons between banks of high buildings. Real physical "selves must be locatable geographically. . . . [A]lthough the virtual self may be disembodied and multiple, it does remain tied to an embodied point" and "breakdowns within virtual societies are tied to self, space and the role of place" (Kolko & Reid, 1998, pp. 224, 226, 228). A community is also "bound by place" and "involves all our senses," and "must be lived" over time, not merely joined as a virtual community that seduces and removes us from localities (Doheny-Farina, 1996, p. 37).

Geographical place also matters for communal values and experience and a sense of community, which requires "human occupancy, commitment, interaction and living among and with others" (Jones, 1997, p. 16). In contrast, the community on the Internet "is in large part incidental to activity that takes place herein," which brings "a sense of connectedness, but it is an aimless connectedness" (p. 17). The virtual lacks the visceral with its human flesh and feeling, taste and touch, sound and smell. That visual communal ritual essential to communication in the grounded "public square" is unlike a virtual, pseudo, or counterfeit site, without the chance, risk, and surprise that is indigenous to urban diversity.

The deterministic technocrats and neo-Luddites have both predicted an abolition of space and time and "blindly support the idea of a perfect relationship between the physical world and cyberspace" (Aurigi, 2005, p. 20); but an amalgam and "re-mediation" has modified early "excited rhetoric" about cyberspace replacing material space (Graham, 2004). The early "heady and unrealistic" literature on the impact of the Internet was "irresponsible" inasmuch as space, geography, and borders have not ended, and new boundaries have been erected in the context of transnational forces and a weakening of the nation-state (Carey, 2005). Places still matter for cities in the unwalled postmetropolis relocated virtually by the quick telecommunications movement of capital to anywhere. New "citi-states" (Pierce, 1995) and a "third wave" of cities are less dependent on one geographical place (Ruchelman, 2000), as are "edge cities" (Garreau, 1991), "world cities" (Moss, 1987; Shatkin, 1998), "global cities," and "global media cities" in an "urban media cluster" where multinational markets and cultures merge in a flow of cultural forms and products (Kratke, 2006).

The displacement of cities is also related to the bodily and psychological impact of mobile communications that may also be emerging as more significant than mere spatial displacement of cognition and community, because the mobile machine "corresponds to deep, primordial human communicational urges" (Nyíri, 2003, p. 12). The new cognitive environment created by mobile phones may be leading to changes in mental architecture (Pléh, 2003; see also Houy, chapter 4 in this volume) and in graphics (Tversky, 2003), with implications for the long-discussed notions of what are organic, real, communal communities, as compared with contractual, societal, informational communities. Like Marshall McLuhan's media extensions of the body, mobile phones become "telephonic cyborgs" in "a world of disembodied sounds—of speech displaced in space and time from its origins . . . conversations without the confrontation of bodies . . . in places that cannot be found on city maps" (Mitchell, 1995, p. 36).

Mobile phone technology extends and mediates the human body like fashion, with "an item of decorative display" related to design that is "strongly connected with ingrained human perceptions of distance, power, status and identity" (Katz & Sugiyama, 2005, p. 63). It is a status symbol (Özcan & Koçak, 2003), like smoking a cigarette is at times (Steinbock, 2005). Children wearing mobile computing and communication devices found that cell phones help them develop a new sense of urban place by observing city geographies of soundscapes that create a virtual digital landscape (Jones, Williams, & Fleuriot, 2003).

The entire body-to-body communication with postures and gestures "intermingles profoundly with forms of mediated communication" in "an increasingly evanescent prototype" of communication more appropriate than the older face-to-face interaction (Fortunati, 2005, p. 53). However, "contextual sensibilities" still prevail at the heart of mobile phone communication, which helps people to be "in touch," and not just to share, convey, or exchange information (Green, 2003). New Internet technologies also are only partially integrated into the metamorphic assimilation and sense of belonging by diverse minorities who may not share the traditional common ethnic geographical community of place (Matei & Ball-Rokeach, 2003).

Whether mobile communication will displace or replace cities has been focused largely on contrasting positive or negative evaluations and projections of the impact of the technology with words like virtual *community* rather than geographic *city*. Forecasts on the technology vary from those who fear it to those who foment its usage. For optimists, the virtual "victory over distance" (Cairncross, 2001) means "new freedoms, rights, cultural identities and authenticity, new creativity, new art forms, new coalitions, new hierarchy, and communicative, dialogue-based personalities and communities" (György, 2003, p. 103). Others are skeptical about what happens to long-held notions of community and society (for example, Georg Simmel, Emile Durkheim, and Ferdinand Tönnies) when interpersonal and institutional relationships are both virtual and mobile, displaced, deterritorialized, and alienated (Green, 2003). This alarm of both fear and hope over the displacement of community and the blurring of place and identity is sounded in regard to the disembodied self and cheapening of personhood. "We now live in cabled enclaves" and "we have forgotten how to locate the 'there,'" but the media technologies "can aid us in the search for community" and "expand community rather than be used by them (media) to replace community" (Bugeja, 2005, pp. xiii, 23, 28, 37). "The Luddite's estrangement is as perilous as the marketer's spam or the guru's infomercial" (p. 112).

The indicted mobile devices "disrupt and diminish the public sphere," colonize public space for individualized, personal interaction, "challenge the decorum of established social settings," "set up a barrier between ourselves and our physical situation" leading to balkanized cliques of like-minded persons in "walled communities," rather than on the "commons," with geographical face-to-face trust and reciprocity, public civility, and shared values (Ling, 2004). They break down the walls and blur the boundaries between public and private, home and work, country and region, with "a sense of location, and a sense of home base becoming lost," and with the "acceleration of the erosion of public-private distinctions," as "public space is destroyed, and even colonized, by private talk that interferes with organic interactions or that prevents public interaction" (Katz & Aakhus, 2002, pp. 9, 15, 18). "Whenever the mobile phone chirps, it alters the traditional nature of public space and the traditional dynamics of private relationships" (p. 301).

A more positive outlook suggests that the mobile phone (like a watch, calendar, or computer) becomes that "city in your pocket" and becomes "a virtual agora" and a "third place" for meeting, although it's a nonplace freed from physical or geographical boundaries (Kopomaa, 2002, pp. 241–245). This allows "living in real time in the same rhythm with someone" in spontaneous interaction with telecompanions, in synchronized living with expanded and increased social interaction and vitality in public space, where phone users "find new ways of attaching themselves to the hub of the city" (p. 245). Urban space becomes "a common living room," a neocommunity of shared interests (Maffesoli, 1996). New notions of the local arise, where time is replaced by continuous access, the future is changing and negotiable, and "communality is not manifested physically as it never acquires a concrete existence" (p. 243). The "placelessness of communication" decentralizes interaction (Fernback, 2005, p. 5). Architects, urban designers, and city planners now face new notions of past, present, and future time as they renew and revitalize cities with new information and communication technologies (ICTs), which have altered the form and structure of cities.

The mobile mediapolis will probably survive in some new form, and in the "real-time city," "mobile communication devices will have a profound effect" and "might change the trajectory of future urban form and function" through "urban metabolism" (Townsend, 2000, pp. 85, 95). Energies and processes of the new media technologies could force the existing cities to accommodate, adjust, and adapt to avoid both their own obsolescence and that of the old media (Gumpert & Drucker, 2006). It has been found that mobile technology can reinforce existing social networks and intimate relationships in urban behavior, rather than replace them with new interactions. Once interaction is freed from fixed locations in space in this "new communications technology regime," phones become like umbilical cords (or pacifiers), "pulling the information society's digital infrastructure into their very bodies" (Steenson, 2006, pp. 1, 9–10).

This appears to take place while people wander and roam like urban nomads and schools of fish with urban "navigational tools" and eventually with telephone numbers rather than geographic locations for people in phone-space. City planners facing this challenge "run the risk of losing touch with the reality of the city streets" inasmuch as modernism must "sustain order and coherence in the face of the acceleration of time and the compression of space" (Townsend, 2000, pp. 101–102). The mobile phone has radically changed the framework of space and time in the city by its uncontrolled appropriation and relocation of a significant physical public whose "nomadic intimacy" of being "present, but absent" (Fortunati, 2002) allows these nomads to "live" in a "mobile home" (Leonini, 1988).

In this context, "computing is not about computers any more. It is about living" (Negroponte, 1996, p. 6). Mobile phones are "machines that become us" and in public space they have become this "absent presence" (Katz, 2003). That combination of the mobile phone with a person creates a fourth form and theory of communication newly defined as "apparatgeist," representing the spirit of the machine (Katz & Aakhus, 2002, p. 301). Also being explored are the spiritual and transcendental uses of mobile communication in religious practices of prayers, evangelistic recruitment, ringtones, the placement of phones in coffins or at grave sites to connect to the afterlife, and the capture of miracles on cameras (Katz, 2006). Inventors are aware that machines are becoming bodies and are trying to create the first mobile "smellophone" that captures playback odor (*London Daily Mail*, 2006). The Walkman has also been examined as an "architect" of privatized electronic narcissism (Chen, 1998), and research hints that excess usage of the cell phone may cause an addiction similar to drugs (McIntyre, 2006).

The movement of the virtual body through teleportation is perhaps the ultimate combination of transportation and communication through quantum cryptography and quantum computing transferring the virtual self without passing through ordinary space and place (Darling, 2005). And beyond that looms the transhuman techniques of bionic transplantation of body organs and computer chips to communicate in the posthuman world through new notions of personhood, where people become mere machines. If microscopic mobile phones are implanted under the skin (as some at Nokia have suggested), and if everyone is accessible all the time, we may then have invented telepathy (Townsend, 2000).

If bodies in the city become mobile machines, and the machine and mobility "[reinforce] this sense of disconnection from space" (Sennett, 1994, p. 19), cities could face even more displacement or even replacement. One concern of skeptics is about the "shift from a definition of personhood embedded in reciprocal, embodied, and co-located relations of mutual obligations and interdependency . . . to an understanding of persons as 'disjointed, interpellated alienated subjects'" (Green, 2003, p. 44). These cyborgs, merged with their cell phones, lack geographical presence because their minds are on the phone and not where their physical bodies are located. People with iPods pay little attention to their immediate surroundings, as they turn public space into private space. They "sing along to their music" while looking directly at other people, but they do not interact with them (Visser, 2005, p. 2). Their virtual video screens achieve technotranscendence whereby they could be at any place in any city at any time (Crawford, 2005).

The city has historically been experienced through physical sensations in space, by how the body moves, hears, smells, eats, dresses, bathes, and makes love; but dull, monotonous modern urban design, planning, and architecture have deprived the senses and created tactile sterility in space. City builders have lost connections to the human body, and tactile reality has been weakened by the geographical shift of people into fragmented spaces, which become a mere functional means for motion and movement as an end in order just to "move through," but not as a means to be aroused (Sennett, 1974, pp. 17–18).

Long before mobile phones, the mass media have divided the lived from the represented urban experience, which has been accelerated by the new mobile geography. When in motion, "the traveler, like the television viewer, experiences the world in narcotic terms; the body moves passively, de-sensitized in space, to destinations set in a fragmented and discontinuous urban geography" (Sennett, 1974, p. 18). The gated, fenced design of modern cities accommodates people's fears of touching bodies during communication. Small wonder that touching pets has developed as a popular cultural form of urban communication (Burd & McKay, 2002). Modern urban form and order have discouraged physical contact, as modern technologies desensitize the body when urban crowds are dispersed to consumption sites like malls. In modern cities, with bodies freed from space, the fear of touching and the threat of the physical presence of others arise in a reduced communal and ritual space, except for in temporary, momentary, and controlled communication sites like sports stadiums. The freely moving modern civic body is usually unaware of other humans, compared with people packed together historically in "old constructions in stone forcing people to touch in the flesh in the urban form of the 'body politic,'" where people were "entirely confident in themselves and at home in their city" (Sennett, 1974, pp. 21–22).

Civic bodies in motion via cell phones may, however, enrich communication behavior in "third places" like restaurants, hotel lobbies, markets, rail stations, and airports (Lasen, 2002), perhaps like the personal, geographic face-to-face social interaction in more physical "third places" (Oldenburg, 1989); in the diverse enclaves threatened by the death of city neighborhoods (Jacobs, 1961); and in the assemblage of Simmel's strangers coming "together alone" in streets, at bus stops, in parking lots, and at bars (Morrill, Snow, & White, 2005). Cities are adjusting to these mobile and wireless communications that are breaking down the city walls (Barrett, Elstrom, & Arnst, 1997), as digital mapping documents the real world and integrates cartography into the Internet, global satellite positioning, and mobile phones (Levy, 2004). A tiny handheld, pocket-sized television aids mobile communication (Baig, 2006), and the cell phone's satellite-based navigation system can tell users where they are located and what they may be searching for or watching on a public street (Markoff & Fackler, 2006).

Beyond the chat rooms of the Internet's "Talk City" (Talk City, 2004), iPod listeners can hear stories of streets as they walk on them (Civic Strategies, 2006). Police have linked city phones to their mobile radios (LookSmart, 1991), and "Boston's first downloadable mayor" has used a podcast to explain the city's gun-buyback program to youth who do not read newspapers (Civic Strategies, 2006). Newspapers can place the full text of local and national news on mobile phones (Burke, 2005); advertisers have used cell phones to reach the streets beyond traditional outdoor advertising (Haskins, 2001); and a national wireless, broadband mobile network, "NexGen City," has been developed to deliver integrated, high-speed data communications systems to cities and counties.

The predicted displacement and radical dislocation of big central city downtowns has not occurred, as they adjust to and integrate with the new media technologies. The mayor of Paris welcomes the wireless Internet as "a decisive tool for international competition," to make Paris "the most connected capital city in the world" (Reuters, 2006). New York post-9/11 as a site for "world trade" still matters as CEOs and corporate headquarters remain centralized and concentrated in Manhattan to "establish a global footprint" for that "global city" (McGeehan, 2000, p. A1). There, "bosses can network face-to-face with their peers (in everyday breakfast meetings and 5-minute breaks) in the hub of the financial, legal, and communications industries, while keeping the rank and file in less expensive quarters in suburbs and other cities" (p. A13). New York's downtown experimental focus on its convergence of mobile and wireless media "spectropolis" involves "wide-ranging experiences from screen-based interactions to audio narratives and physical installations" at eight hotspots that include "a wireless bicycle that chalks text messages from the Web into city streets, a portable hearing device that transmits a soundscape responding to immediate surroundings, and a computer station for free download of digital art objects" (AllBusiness, 2004). Mobile communications media in "mediapolis" continue to change the urban experience in public and private space, but the city as metaphor and place survives. The city does not disappear, but is rearranged and relocated as it adapts, adopts, and augments new media technologies instead of being replaced by media or communication—its historic reason for existence.

References

Adams, P. C. (1997). Cyberspace and virtual places. *Geographical Review*, 87(2), 155–171.

AllBusiness. (2004). Communication technology is talk of downtown. Retrieved from http://www.allbusiness.com/operations/facilities-commercial-real-estate/230899–1.html.

Augé, M. (1995). *Non-places: Introduction to an anthropology of supermodernity*. London: Verso.

Aurigi, A. (2005). *Making the digital city: The early shaping of urban Internet space*. Hants: Ashgate.

Baig, E. C. (2006, August 18). Will consumers tune in to a tiny TV in their hand? *USA Today*, pp. B1–B2.

Barrett, A., Elstrom P., & Arnst, C. (1997, January 20). Vaulting the walls with wireless. *Business Week*, 45–46. Retrieved October 22, 2006, from http://www.businessweek.com/archives/1997/b3510092.arc.htm?chan=search.

Bugeja, M. (2005). *Interpersonal divide: The search for community in a technological age*. New York: Oxford University Press.

Burd, G. (2002, August). The search for natural regional space to claim and name built urban place. Paper presented at the meeting of the American Sociological Association, Chicago, IL.

Burd, G., & McKay, M. (2002, August). Pets as urban communication partners: Touching as tactile "talk" in cities. Paper presented at the meeting of the American Sociological Association, Chicago, IL.

Burke, P. (2005, October). Cox newspapers reach new mobile audience. *COXnet*, p. 26.

Cairncross, F. (2001). *The death of distance: How the communication revolution is changing our lives*. Boston, MA: Harvard Business School Press.

Carey, J. (2005). Historical pragmatism and the Internet. *New Media and Society*, 7(40), 443–455.

Chen, S-L. S. (1998). Electronic narcissism: College students' experiences of Walkman listening. *Qualitative Sociology*, 21(3), 255–275.

Civic Strategies E-Letter. (2006, July). If these streets could talk: Welcome to a new era. Retrieved from otwhite@civic-strategies.com.

Crawford. W. (2005, May). *EContent*, 28(5), 43. Retrieved October 23, 2006, from http://www.econtentmag.com/Articles/ArticleReader.aspx?ArticleID=7944&AuthorID=11.

Darling, D. (2005). *Teleportation: The impossible leap*. Hoboken, NJ: John Wiley & Sons.

de Sola Pool, I. (1977). *The social impact of the telephone*. Cambridge, MA: MIT Press.

Doheny-Farina, S. (1996). *The wired neighborhood*. New Haven, CT: Yale University Press.

Dreier, P., Mollenkopf, J., & Swanstrom, T. (2001). *Place matters: Metropolitics for the twenty-first century*. Lawrence: University Press of Kansas.

Drucker, S. (2006, April). Wi Fi and Philadelphia. Paper presented at the meeting of the Eastern Communication Association, Philadelphia, PA.

Drucker, S., & Gumpert, G. (1991). Public space and communication: The zoning of public interaction. *Communication Theory*, 1(4), 294–310.

Drzewiecka, J. A., & Nakayama, T. K. (1998). City sites: Postmodern urban space and the communication of identity. *Southern Communications Journal*, 64(1), 20–31.

Fernback, J. (2005, April). Information technology and community networks: A symbolic interactionist perspective on urban regeneration. Paper presented at the meeting of the International Communication Association, New York.

Fortunati, L. (2002). The mobile phone: Towards new categories and social relations. *Information Communication and Society*, 5(4), 513–528.

———. (2005). Is body-to-body communication still the prototype? *The Information Society*, 21(1), 53–61.

Garreau, J. (1991). *Edge city: Life on the new frontier*. New York: Doubleday.

Geser, H. (2004). Towards a sociological theory of the mobile phone. Retrieved from http://socio.ch/mobile/t_geser1.pdf.

Graham, S. (2004). Beyond the "dazzling light": From dreams of transcendence to the "remediation" of urban life. *New Media & Society*, 16(1), 16–25.

Green, N. (2003). Community redefined: Privacy and accountability. In K. Nyíri (Ed.), *Mobile communication: Essays on cognition and community* (pp. 43–55). Vienna: Passagen Verlag.

Gumpert, G., & Drucker, S. (2006). Obsolescent media and the urban landscape. *InterMedia*, 34(2), 31–35.

György, P. (2003). Virtual distance. In K. Nyíri (Ed.), *Mobile communication: Essays on cognition and community* (pp. 97–103). Vienna: Passagen Verlag.

Haskins, W. (2001, April 24). Take it to the streets. *PC Magazine*, 28(8). Retrieved from http://www.findarticles.com/p/articles/mi_zdpcm/is_200104/ai_ziff3590.

Henricks, M. (2002, May). Flexin' the city: Think you know the face of urban America? The U.S. city is in for a workout. *Entrepreneur Magazine*, 30, 13.

Jacobs, J. (1961). *The death and life of great American cities*. New York: Random House.

Jassem, H. (2005, November). Municipal wi-fi-ing of the United States. Paper presented to the meeting of the National Communication Association, Boston, MA.

Jones, O., Williams, M., & Fleuriot, C. (2003). A new sense of place? Mobile "wearable" information and communications technology devices and the geographies of urban childhood. *Children's Geographies*, 1(2), 165–180.

Jones, S. G. (1997). The Internet and its social landscape. In S. G. Jones (Ed.), *Virtual culture: Identity and communication in cybersociety* (pp. 7-35). London: Sage.

Katz, J. E. (2003). A nation of ghosts? Choreography of mobile communication in public spaces. In K. Nyíri (Ed.), *Mobile democracy: Essays on society, self and politics* (pp. 21–31). Vienna: Passagen Verlag.

———. (2006). *Magic in the air: Mobile communication and the transformation of social life*. New Brunswick, NJ: Transaction.

Katz, J. E., & Aakhus, M. (Eds.). (2002). *Perpetual contact: Mobile communication, private talk, public performance*. Cambridge: Cambridge University Press.

Katz, J. E., & Sugiyama, S. (2005). Mobile phones as fashion statements: The co-creation of mobile communication's public meaning. In R. Ling & P. E. Pedersen (Eds.), *Mobile communications: Re-negotiation of the social sphere* (pp. 63–81). London: Springer-Verlag.

Kolko, K., & Reid, E. (1998). Dissolution and fragmentation in online communities. In S. G. Jones (Ed.), *CyberSociety 2.0: Revisiting computer-mediated communication and community* (pp. 212–229). Thousand Oaks, CA: Sage.

Kopomaa, T. (2000). *The city in your pocket: Birth of the mobile information society*. Oxford: Oxford University Press. Retrieved from http://www.tkk.fi/Yksikkot/YTK/julkaisu/mobile.

———. (2002). Mobile phones, place-centred communication and neo-community. *Interface*, 241–245. Retrieved from http://www.casa.ucl.ac.uk/cyberspace/kopomaa_mobile_phones.pdf.

Kratke, S. (2006). Global media cities: Major nodes of globalizing culture and media industries. In N. Brenner & R. Keil (Eds.), *The global cities reader* (pp. 325–331). London: Routledge.

Lasen, A. (2002). *The social shaping of mobile and fixed networks: A historical comparison*. Surrey: Digital World Research Centre, University of Surrey.

Leonini, L. (1988). *L'identita smarrita*. Bologna: Mulino.

Leung, L., & Wei, R. (2000). More than just talk on the move: Uses and gratifications of the cellular phone. *Journalism and Mass Communication Quarterly*, 77, 308–320.

Levy, S. (2004, June 7). NextFrontiers: Something in the air. *Newsweek*, 45–68, 76.

Ling, R. (2004). *The mobile connection: The cell phone's impact on society*. San Francisco, CA: Morgan Kaufmann.

London Daily Mail. (2006, June 29). Smellophone with a nose of its own. Retrieved from http://www.dailymail.co.uk/pages/live/articles/news/news.html?in_article_id=393075&in_page_id=1770.

Long, N. (1972). *The un-walled city: Reconstituting the urban community*. New York: Basic Books.

LookSmart. (1991, September). Gainesville, Georgia, links city phones with mobile radio. Retrieved from http://findarticles.com/p/articles/mi_m0CMN/is_n9_v28/ai_11238709.

Lynch, K. (1972). *What time is this place?* Cambridge, MA: MIT Press.

Maffesoli, M. (1996). *The contemplation of the world: Figure and community style*. Minneapolis: University of Minnesota Press.

Markoff, J., & Fackler, M. (2006, June 28). With a cellphone as my guide. *New York Times*, pp. C1, C7.

Marshall, A. (2000). *How cities work: Suburbs, sprawl and the roads not taken*. Austin: University of Texas Press.

Martindale, D. (1958). Prefatory remarks. In M. Weber (D. Martindale & G. Neuwirth, Trans. & Eds.), *The city* (pp. 34–35). New York: Collier.

Matei, S., & Ball-Rokeach, S. (2003). The Internet in the communication infrastructure of urban residential communities: Macro- or mesolinkage. *Journal of Communication*, 53(4), 642–657. Retrieved from http://www.cios.org/getfile/MATEI_V11N201.

McGeehan, A. (2000, July 3). Once again, the boss is in at the New York office. *New York Times*, pp. A1, A13.

McIntyre, S. (2006, August 21). BlackBerry addiction "similar to drugs." *London Daily Mail*. Retrieved from http://www.dailymail.co.uk/pages/live/articles/news/news.html?in_article_id=401646&-in_page_id=1770.

Mitchell, W. J. (1995). *City of bits: Space, place and the infobahn*. Cambridge, MA: MIT Press.

Morrill, C., Snow, D., & White, C. W. (2005). *Together alone: Personal relationships in public places*. Berkeley: University of California Press.

Morris, D. (2005). *It's a sprawl world, after all*. Gabriola Island, British Columbia: New Society.

Mosco, V. (2004). *The digital sublime: Myth, power, and cyberspace*. Cambridge, MA: MIT Press.

Moss, M. L. (1987). Telecommunications, world cities and urban policy. *Urban Studies*, 24, 334–346.

Negroponte, N. (1996). *Being digital*. New York: Vintage (Random House).

Nyíri, K. (2003). Introduction. In K. Nyíri (Ed.), *Mobile communication: Essays on cognition and community* (pp. 11–23). Vienna: Passagen Verlag.

O'Keefe, G. J., & Sulanowski, B. K. (1995). More than just talk: Uses, gratifications, and the telephone. *Journalism & Mass Communication Quarterly*, 72(4), 922–993.

Oldenburg, R. (1989). *The great good place: Cafes, coffee shops, community centers, beauty parlors, general stores, bars, hangouts and how they get you through the day*. New York: Paragon House.

Özcan, Y. Z., & Koçak, A. (2003). Research note: A need for a status symbol? Use of cellular telephones in Turkey. *European Journal of Communication*, 28(2), 241–254.

Pascal, A. (1987). The vanishing city. *Urban Studies*, 24, 597–603.

Pierce, N. R. (1995). *Citi-states: How urban America can prosper in a competitive world*. Washington, D.C.: Seven Locks.

Pléh, C. (2003). Communication patterns and cognitive architectures. In K. Nyíri (Ed.), *Mobile communication: Essays on cognition and community* (pp. 127–141). Vienna: Passagen Verlag.

Reuters. (2006, July 4). Paris wants wireless Internet access across city. Retrieved from http://freepress.net/news/16387.

Ruchelman, L. I. (2000). *Cities in the third wave*. Chicago, IL: Burnham.

Rule, J. B. (2002). From mass society to perpetual contact: Models of communications technologies in social content. In J. E. Katz & M. A. Aakhus (Eds.), *Perpetual contact: Mobile communication, private talk, public performance* (pp. 242–254). Cambridge: Cambridge University Press.

Sennett, R. (1974). *The fall of public man*. New York: Vintage (Random House).

———. (1994). *Flesh and stone: The body and the city in Western civilization*. New York: W. W. Norton.

Shatkin, G. (1998). "Fourth world" cities in the global economy: The case of Phnom Pen, Cambodia. *International Journal of Urban and Regional Research*, 22(3), 378–393.

Simmel, G. (1950). The metropolis and mental life. In K. H. Wolff (Trans.), *The sociology of Georg Simmel* (pp. 409–424). New York: Free Press. (Original work published 1903).

Steenson, M. W. (2006). *The excitable crowd: Characterizing social, mobile space*. Unpublished master's thesis, Yale University, New Haven, CT.

Steinbock, D. (2005). Seeing is believing. *Communication World*, 22(6), 19–21.

Steinhardt, N. S. (2000). Representations of Chinese walled cities in the pictorial and graphic arts. In J. D. Tracy (Ed.), *City walls: The urban enceinte in global perspective* (pp. 419–460). Cambridge: Cambridge University Press.

Talk City. (2004). Looking for chat rooms? Retrieved from http://www.talkcity.com/homepage.htm?flash=y.

Townsend, A. M. (2000). Life in the real-time city: Mobile telephone and urban metabolism. *Journal of Urban Technology*, 7(2), 85–104.

Tracy, J. D. (Ed.). (2000). *City walls: The urban enceinte in global perspective*. Cambridge: Cambridge University Press.

Tversky, B. (2003). Some ways graphics communicate. In K. Nyíri (Ed.), *Mobile communication: Essays on cognition and community* (pp. 143–156). Vienna: Passagen Verlag.

Visser, G. (2005, October 11). iPods, mobile phones and community. Retrieved from http://www.smartmobs.com/archive/2005/10/11/ipods_mobile_p.html.

Wellman, B. (2001). Physical place and cyberspace: The rise of personalized networking. *International Journal of Urban and Regional Research*, 25(2), 227–252.

———. (2005). Connecting community: On- and offline. Retrieved from http://www.chass.utoronto.ca/~wellman/publications/contexts/contexts-3a.htm.

Wellman, B., & Gulia, M. (1996). Net surfers don't ride alone: Virtual communities as communities. Retrieved from http://www.acm.org/~ccp/references/wellman/wellman.html.

·4·

LIVING AND LOVING IN THE METRO/ELECTRO POLIS

Understanding the Neurobiology of Attachments in a Society with Ubiquitous Mobile Information and Communication Technologies

Yvonne Houy

"Human connections shape the neural connections from which the mind emerges" (Siegel, 1999, p. 2). This short sentence has profound implications for a society in which mobile technologies are used frequently in interpersonal relationships. Daniel Siegel, psychiatrist and founder of the field of interpersonal neurobiology, is summarizing groundbreaking work on the neurobiology of interpersonal relationships. What he means is that a connection between minds can only find its fullest expression through physical proximity. Human attachments are created through the communication of emotions that occurs primarily through nonverbal means, such as tone of voice and facial and body movements (Schore, 2001; Siegel, 1999, 2001, 2003). The subtleties of nonverbal signals cannot be translated through current mobile or other computer-mediated communication (CMC) technologies. If physical proximity to loved ones is not only desirable but necessary for emotional and physical health, what are the implications for the emotional and physical health of people who routinely displace place, frequently interacting not with those who are physically close to them, but instead with others far away, via mobile phone or laptop computer? The suggestion that personal mobile devices necessarily impede meaningful interactions and potentially stunt emotional health—recently made by a book popularizing neurobiological research (Goleman, 2006)—is a Luddite-sounding reaction that deserves examination.

Although the emerging field of interpersonal neurobiology has not yet explored this question, recent findings suggest that humans should not be able to thrive emotionally in the virtual spaces of the "Electropolis." However, Electropolis is a "place" in which personal interaction does succor strong emotional ties that can trigger strong emotional responses, according to research on online communication. Can mobile communication technologies (MCTs) foster emotional attachments because of their potential for increased interactions, or could the reduction of emotional cues in mediated communications interfere with emotional attachments?

Current Research on Online Social Life

Communication using mobile communication devices has some similarities to online communication—both are mediated communication at a distance, and both reduce social and emotional cues—but research about the social implications of mobile devices is not as well established, and much of our current understanding of the social implications of mobile technologies derives from research about online social life. Interpersonal relationships conducted using mobile technologies are not the same as relationships conducted online, but there are significant overlaps. Online communication can involve mobile devices, such as laptops with wireless modems, or mobile communicators, such as the BlackBerry, but it can also involve less mobile devices, such as desktop computers using wired Internet access. Research on online social life is thus of value, but it cannot answer key questions about the current proliferation of mobile communication devices.

Research on the social aspects of CMC has come a long way from the mid-1980's skepticism about whether online social life could ever be a major part of using CMC, to an understanding in the mid-1990s that the opportunities for interpersonal communication are among the most important features of CMC (Baym, 2006). Much of the research on social life using CMC devices assumes, in obvious and subtle ways, that face-to-face interaction is "real" interaction. After more than twenty years of CMC research, social life through online communication is "officially" no longer seen as strange or aberrant; after all, individuals have conducted relationships with others via communication technologies since the advent of writing millennia ago. However, the underlying questions that guide much of the research on personal communication media, such as the Internet and MCTs, are: How can or do online users conduct social relationships at a distance? and What is the relationship between face-to-face interaction and interaction through technologically mediated means? (See Baym, 2006, for an overview of research on online social life.) These underlying questions are posed as if it were surprising to create and maintain relationships while separated over distance.

It does not help researchers that theories of personal relationships tend to be biased toward face-to-face communication, and define relationships through face-to-face characteristics. This bias impedes understanding of how relationships develop in CMC (Lea & Spears, 1995). CMC social life researcher Nancy Baym (2006) summarizes the continuing skepticism: "despite the implications of many interpersonal and postmodern theories that people can't or won't form personal relationships through CMC, people do, and do so often and fairly successfully. CMC, and the Internet, offer new opportunities for creating relationships" (p. 43). Her research on online relationships via Usenet groups confirms this (Baym, 2000). We know from such research, and through personal experience, that relationships are formed and maintained using mediated communication technologies. The increasing use of mobile technologies is largely a continuation of the online phenomenon that has been researched extensively. This bias toward face-to-face communication should not be surprising, because face-to-face communication has been the primary form of interpersonal communication in human history.

While interactions mediated through communication technologies are nothing new, today's information and communication technologies (ICTs), especially mobile ICTs, allow their users to conduct relationships in qualitatively different ways than have been possible before. Letters have kept friends and families in touch over long distances—at least those who had the skills to write them and resources to send them. Physically conveying these written messages took time, and the physical distance more or less correlated with the time it took to have the message received and get a response. Over the past century and a half, the telegraph and then the telephone have made the world seem "smaller," as far-flung travelers—at least those with the financial means to take advantage of these technologies—could dialogue, with minimal time lag, with loved ones left behind. What makes the communication technologies available today different is their mobility. Sender and receiver are no longer dependent on the fixed physical locations of post offices, telegraph stations, and telephone landlines, and the phones and Internet jacks fixed onto them; sender and receiver can take their personal communication devices with them and communicate directly with each other by voice or in writing from wherever their personal mobile communication devices are in service.

Over the past decade, this potential for near-instantaneous accessibility almost anywhere has changed how "information-haves" interact socially. We have all seen people sitting in cafés not interacting with those across from them, but with others, who knows where, via cell phone. And we have all seen people emailing on their wireless laptops rather than paying full attention to the meeting or lecture they were attending at the same time. As the title of this volume so succinctly says, mobile technologies do displace place in multiple ways. In doing so, they also displace physical proximity and face-to-face interaction in intimate situations. Our friends and loved ones are potentially only a phone call away, no matter where they might be at the moment—at least while in range of wireless signals. Increasingly, information-haves can have their bodies located in their physical world—symbolized by the term *Metropolis*—while their minds are in Electropolis as they communicate with physically distant others through their personal mobile communication devices. In so doing, they are creating, not just populating, the new hybrid space of the Metro/Electro Polis (Houy, 2005).

The emotional implications of living in this hybrid space of the Metro/Electro Polis is being explored, but we have as of yet more questions than answers. Can humans live fulfilled emotional lives in the Electropolis? What does living in the hybrid space of the Metro/Electro Polis mean for humans' emotional life? What aspects of the physical face-to-face world of the Metropolis are necessary for full emotional interaction? There are some preliminary answers to these questions in research on Internet technologies, but much less research on the impacts of mobile ICTs specifically.

In the literature about relationships that are conducted using online and/or MCTs, there is a notable lack of research that uses the established psychological concept of "attachment." The nature and function of attachment between loved ones has been researched in psychology since the 1950s, with significant refinements in the 1990s in the understanding of the role of the mind and body in attachment. New findings in human neurobiology have uncovered the neurological basis of human attachments, and research on the physiological effects of affect in close relationships is increasing our understanding of the importance of attachment for physical, emotional, and mental health.

This continuously evolving and rapidly expanding field of knowledge enables researchers to focus on the effects on individual users. Baym (2006) has lamented that most research into CMC has focused on groups and averaged out individuals. This is a limiting approach that is bound to impede understanding: CMC has had such a great variety of effects on users that a recent study concluded that it is impossible to talk about one main effect on an average user (McKenna & Bargh, 2000). In particular, individual differences in the role CMC plays in the maintenance of existing relationships should receive "considerably more attention" (Baym, 2006, p. 45). In addition, the relationship between unmediated interaction and mediated interactions needs to be examined further, because research has shown that, for many partners, CMC became just one way to interact (Wellmann & Gulia, 1999). These critiques of current research should also be raised about research into social interactions with MCTs, because of their close relationship to CMC technologies.

Attachment Theory

This section summarizes the findings in attachment research because of its relevance to understanding the interactions of individuals in close relationships. The work of John Bowlby (1973, 1979, 1980, 1988) forms the fundamental basis of the current understanding of attachment. His investigation of infants' attachment system postulated that infants' brains are hardwired to seek physical or psychological contact with their caregivers. Attachment behaviors, such as visually tracking, following, and in other ways promoting physical contact with their primary caregiver, including protesting by crying when the caregiver leaves, probably evolved to ensure that caregivers remained near infants and thus increased infants' chances for survival. By the end of the first year, primary caregivers form the center of a child's social life. Infants form attachments even in cases of child neglect and abuse, but these bonds lead to problematic attachment styles.

Following Bowlby's work, Mary Ainsworth (Ainsworth, 1972, 1990; Ainsworth et al., 1978) researched how different caregiving styles affected the attachment behavior of small children. Specifically, Ainsworth looked at how twelve- to eighteen-month-old children reacted to their primary caregiver in a stressful situation, called the Strange Situation. Toddlers were left alone by their primary caregivers (usually mothers) in a room that was new to them. Usually toddlers looked for and then protested when they realized that their attachment figure was not available. The kind of attachment that caregiver and child shared was revealed by how the child acted when the caregiver returned. In a secure caregiver-child attachment, the child would seek physical proximity with their caregiver once he or she returned, crying or protesting only briefly, but then calm quickly and resume exploring the new environment. Insecure attachments between caregiver and child took two forms, according to Ainsworth: anxious-ambivalent and avoidant. Children who had mixed reactions to their returning caregiver, perhaps by approaching and then avoiding contact with the caregiver, or who remained agitated and did not resume normal activities, were said to have anxious-ambivalent attachments. In avoidant attachments, the child disregarded and avoided the returning caregiver and showed signs of emotional disengagement.

Ainsworth's research shows that different attachment styles in infants and children are related to the caregiver's typical responses to an infant's needs. Secure attachments, sometimes called normal attachments, are the result of sensitive, responsive caregiving. When an infant seeks proximity or communicates a need, a sensitive caregiver rapidly responds positively, and effectively meets the infant's needs. Interacting with an affectionate, consistent, responsive caregiver allows a child to develop a powerful emotional bond that becomes the basis for future emotionally fulfilling relationships. In such normal secure attachments, the original attachment figure becomes a secure base from which a child, and later adult, launches herself into a less secure world.

Less responsive caregiving can lead to the two types of insecure attachment styles, anxious-ambivalent and avoidant. In caregiver-child attachments that show anxious-ambivalent tendencies, the caregiver is not as adept at reading the child's cues of distress, or is less effective in meeting the child's needs, or is inconsistent in meeting the child's needs. These behaviors increase the child's anxiety about whether his or her needs will be met. In avoidant attachments, the child does not expect to be comforted by the caregiver, or, in extreme cases, expects punishment from the caregiver, and therefore does not seek out the caregiver when in distress. Child neglect or abuse typically leads to one of these types of attachments, but even less extreme unresponsive or insensitive caregiving can lead to such insecure attachment styles.

Between about twelve and eighteen months, the developing child internalizes the nature of his or her original experiences as a "working model" of attachments in his or her psychological structures. In response to repeated interactions with significant caregivers, a child develops expectations of what is likely to occur in certain situations. This internalized working model answers in particular the question: How do other people respond to me in times of stress or need?

Attachment Theory and Neurobiology

Neuroscience, in particular neurobiology, is beginning to understand how these internalized working models are established in the mind. Neurobiology traces the arousal and activation of neurons, or flow of energy within brains. Neurophysiological processes within the brain can be traced through brain imaging, assessing metabolic processes or patterns of blood flow, and electroencephalograms (EEGs), assessing electrical activity on the brain surface. Information is processed by the brain as the sensory system responds to stimuli and then represents this as patterns of neural firing, or flows of energy. Different parts of the brain then process this pattern into a coherent whole image or narrative.

Neurophysiologically, the internal representation of attachment and attachment figures occurs because of the repetition of experiences over the first twelve to eighteen months of life. As an infant interacts with his or her primary caregivers, each interaction activates neurons in the brain. As certain kinds of interactions repeatedly fire particular neurons in the same or similar pathways, these pathways etch into the structure of the brain and become more ingrained with each recurrence (Lewis, Amini, & Lannon, 2001; Siegel, 1999, 2001, 2003). In particular, the infant's right brain hemisphere creates connections to the limbic and autonomic nervous system, which is responsible for stress response. Positive interactions with responsive caregivers thus allow an infant to develop coping mechanisms for reacting to new experiences and stress (Schore, 2001). Familiar events—for example, the caregiver's typical response to an infant signaling hunger or wish for connection and affection—become the expected response for the child over time because they are literally etched into the structures of the brain.

There are three fundamentals in the neurobiology of interpersonal experience, a field that seeks to understand how the mind develops through interpersonal relationships. First, the structure and functions of the human brain develop through experiences, especially interpersonal ones. Second, the mind develops through the interaction of interpersonal experiences and neurophysiological processes, and third, the mind is created by patterns of energy flows and the processing of information within individual brains and between the brains of individuals who are relating to each other (Siegel, 1999). In other words, experiences shape the innate, genetically determined development of the nervous system (Lewis et al., 2001; Schore, 2001; Siegel, 1999, 2001, 2003; Solomon & Siegel, 2003).

Thus, early experiences have profound physiological, emotional, and mental effects on individuals throughout their lives. People respond consciously and unconsciously, mentally and physically, to these experience-driven internalized working models of attachment. W. Steven Rholes and Jeffry A. Simpson (2004) summarize the extensive psychological and physical effects of this internalized representation:

> Working models orchestrate behavior, cognition, and affect in close relationships, providing guidance about how to behave, what should be expected or anticipated, and how to interpret the meaning of ambiguous interpersonal events. Working models control attention to and memory for information associated with attachment-relevant events, and they regulate affect—especially negative affect—when attachment-relevant stressors are encountered. (p. 7)

In other words, internal working models affect the whole body and mind.

The plasticity of the brain after childhood is currently being debated. Some argue that the structure of the brain is set after puberty, with important neurological settings occurring in infancy and childhood (Schore, 2001). Others, like Siegel (1999, 2001, 2003), argue that the structure of the brain can continue to change, although early childhood is a crucial time of development in which the very structure of the brain is constructed through interpersonal interactions. The structure of the brain can continue to change throughout a person's life, albeit considerably more slowly and less radically than in early childhood. Empirical evidence in clinical psychology supports Siegel's arguments: internalized attachment styles can change with new experiences, such as new types of relationships, including psychologist-client relationships or other close attachment relationships with different behavioral patterns than those associated with the early childhood caregivers (Siegel, 1999, 2001, 2003; Solomon & Siegel, 2003).

The type of attachment a child had with his or her primary caregiver is likely to mirror his or her attachment style in adulthood, although the style can change because of the plasticity of the brain. An individual carries expectations of likely outcomes in interpersonal interactions into adulthood, but as an individual has different relationships these working models can change. It is more likely, though, that new experiences are assimilated into a particular model than that the model itself changes in response to new interactions and relationships. Thus, adults do not enter relationships with blank slates; they bring with them a history of memories and expectations of how relationships work that tends to correspond with their attachment history (Collins et al., 2004). Each individual has a particular "attachment style"; that is, there are individual differences in tendency to seek emotional ties and presumptions about availability and responsiveness of attachment figures in times of need (Rholes & Simpson, 2004).

The internalized model of normal attachment is emotionally and physically useful in the inevitable increasing separations between child and primary caregiver. The caregiver does not need to be physically available constantly because the relationship continues in the mind of the child. The feeling of having a secure base allows the securely attached child to venture into the world with fewer anxieties than the avoidantly or anxious-ambivalently attached child.

Attachment and Mobile Communication Technologies

The internalization of attachment figures after the first year of life is the feature of human psychology that is the likely reason why humans can sustain relationships via mediated means. The writer or caller remains literally present in the mind of the receiver—because he or she is etched into the neural pathways of the brain—even if that beloved person is not physically present. However, it seems that mediated communication can only approximate the full effect of face-to-face communication. Recent research in neuro- and psychophysiology suggests that humans need close physical proximity to attachment figures for emotional and even physiological well-being.

While there is no definitive definition of adult attachment (Hazan, Gur-Yaish, & Rholes, 2004), the effects of attachments on individuals are becoming better understood. Research on adult attachment is based on the assumption "that it occurs and is manifested at multiple levels, including behavior, cognition, physiology, and emotion" (Hazan et al., 2004, p. 56). Animal research has been significant in our understanding of the physiological effects of attachment on humans, but as of yet there have been no systematic investigations of the physiological aspects of attachment in humans (Diamond, 2001). However, even the scattered evidence in research so far suggests that humans need physical proximity to attachment figures for their short- and long-term well-being.

We do know that affect regulation and physical co-regulation are important aspects of attachments. Indeed, they are thought to be markers of attachment (Hazan et al., 2004). Loved ones can be a calming influence in times of stress, a kind of buffer, and can help regulate our endocrinal and circulation systems just by their physical proximity. For example, research has shown that people have lower blood pressure when they are with romantic partners than when they are with friends (Gump et al., 2001). Because of the human capacity for mental representation, which is related to the internalization of attachments, co-regulation can become increasingly internalized (Hofer, 1984, 1987, 1994, 2005), suggesting that co-regulation could extend beyond immediate interactions (Hazan et al., 2004). This might be yet another reason why human beings can continue to have close relationships via mediated means.

The physiological response to an attachment figure depends on an individual's attachment style, which in turn might influence how an individual uses mediated communication devices to communicate with loved ones. In normal, securely attached individuals, positive echoes of the physiological and emotional effects of close proximity to attachment figures continue, while possible negative physiological effects might plague insecurely attached individuals. In one study, women who were avoidantly attached actually experienced stress when with a romantic partner, while securely attached women did not experience more stress whether they were with their partners or not (Carpenter & Kirkpatrick, 1996).

Because of these physiological responses to close proximity, those with avoidant attachment patterns might use communication devices to keep people at a distance. For someone with an avoidant attachment style, talking on a cell phone rather than face-to-face might end up feeling better. Mobile technologies might enable—for better or worse—those with insecure attachment styles to maintain intimate relationships because they can make themselves verbally and mentally available to their partner while maintaining physical distance.

It might be difficult to distinguish between people who use MCTs to isolate themselves and let these technologies interfere with their face-to-face social lives and people who would be even more isolated without these technologies, as one study of Internet use suggested, arguing for more research into the area (Mitchell, Becker-Blease, & Finkelhor, 2005). Communication technologies might allow shy or lonely individuals to reach out more. One study found positive relationships between loneliness and Internet use in a sample of college students (Matsuba, 2006). Another study found that shyness correlated with involvement in online relationships (Ward & Tracey, 2004). Often the boundary between personally enhancing and problematic use is perception: for some people, pursuing relationships via mediated means might be a move into an escapist fantasy, while for others it is a healthy adaptive strategy to overcome isolation imposed by mental or physical challenges (Fox, 2000).

However, even for those individuals whose MCT use is a net positive for their social lives, something vital is missing when they are engaged in something other than face-to-face communication, because any mediated communication can only approximate the full effects of emotional communication between individuals in physical proximity to each other. Any interpersonal communication, but especially the communication of emotions, occurs not only through the manipulation of written and aural symbols. Siegel (1999) summarizes the difficulty of using words alone to convey the complexity of emotions:

> Complex neural/bodily aspects of emotional processes are not easily translated into words. Nonverbal expressions, including those of the face, tone of voice, and gestures, can transfer information about internal states more fully to the outside world than words can do. . . . Linguistic representations, such as the words "sad" or "angry," are quite limited and distant symbolic packets we send to each other in response to the query, "How are you feeling?" The message is in the medium of how we respond, not in the words alone. (p. 150)

Focusing on symbolic communication, such as spoken or written words, shortchanges the richness of how we interact, but that is what CMC research has examined.

Emotional communication is a profoundly physical act that involves a wide range of nonverbal cues that draw on all senses. Body language, facial expression, timing of responses, eye gaze, and hormones secreted through the skin and released as smell all play a significant role in how humans understand one another (Iacoboni, 2005; Siegel, 1999, 2001, 2003). Conscious facial expressions and body movements are culturally determined, but human beings can control only some physical signs of emotional responses. Internal emotional states are expressed externally through the complex and sensitive muscle endings in the face, which are activated unconsciously through the nervous system. Muscles and the nervous system allow humans and other primates to produce many types of facial expressions, which are subtle and can change rapidly. Human beings interpret these communication cues consciously and unconsciously (Siegel, 1999).

Neuroscientists have discovered that most individuals can actually experience another's emotions when they intentionally attune themselves to nonverbal signals. Humans and primates have so-called mirror neurons that are activated when the human or animal performs a particular act or even watches someone else perform the act (Gallese, Keysers, & Rizzolatti, 2004; Iacoboni, 2005; Iacoboni et al., 2005). These neurons are the reason why people wince when they see someone get hurt. They are the likely basis for empathy between sentient beings. What is startling about these neurons is the extent to which they provide us with insight into others' minds: "Social cognition is not only thinking about the contents of someone else's mind. Our brains, and those of other primates, appear to have developed a basic functional mechanism, a mirror mechanism, which gives us an experiential insight into other minds" (Gallese et al., 2004, p. 403). This experiential insight is what Vittorio Gallese (2004) calls an "embodied simulation producing a shared body state." This shared body state literally merges two discrete physiologies into one connected unit as one person influences another (Diamond & Aspinwall, 2003a, 2003b).

Experiential insight into another's experience—the shared bodily state—is crucial for humans because it allows us to "read" others' attentional focus, judgment of events, and intentions by being able to experience them, and it allows us to understand interactions and to anticipate others' behavior. Communicating and perceiving emotional states probably had an important evolutionary function, because accurate perception allows humans to interpret intentions, and empathy among people in a group establishes the basis for the social life that is crucial for the physical, emotional, and mental survival of a social species. Emotional communication is social communication (Siegel, 1999).

Siegel (1999) calls this ability to perceive another's state of mind by perceiving nonverbal cues in the body and face "mindsight." Not everyone has this ability; some individuals cannot perceive others' emotional states. This deficit may be due to emotional dissociation because of attachment trauma from early childhood or damage to certain brain structures (Siegel, 1999).

Expressing emotions and having others respond in kind is vitally important to the emotional and physical health of an individual, and it allows human brains to develop normally. Seeking proximity to a caregiver and face-to-face communication with eye gazing—the actions that make emotional communication possible—are hardwired from birth for a good reason. Research into the neurobiology of interpersonal experience has found that responsive caregiving is related to the quality of the parent-child attunement of emotional states, which enables the caregiver to respond sensitively and effectively. Infants learn to express and regulate their emotions by tuning in to their primary caregiver's emotional states. The attunement of minds between infant and caregiver in normal secure relationships helps integrate brain functions in an infant's developing mind (Schore, 2001; Siegel, 1999, 2001).

The importance of emotional connection does not stop after infancy and childhood. Siegel (1999) summarizes the importance of emotional communication to adults: "as adults, we need not only to be understood and cared about, but to have another individual simultaneously experience a state of mind similar to our own" (p. 22). Emotions and how they are communicated verbally and nonverbally are fundamental to how humans create and maintain connections. Emotions directly influence the functions of the brain and the entire body, including reasoning and bodily functions (Siegel, 1999). Neurologists interested in the interpersonal implications of their research see the communication of emotions as the central function of communication among human beings.

With current MCTs, however, some of those vital nonverbal cues are filtered out, limiting the ability to communicate emotions. Which cues are filtered out depend on the specific technology. Cell phones, like telephones, can transmit a limited range of voice cues that make emotional states perceivable. Video conferencing has similar issues with transmitting voice cues, but it has the ability to transmit some visual cues, such as unsubtle facial expression and body language, depending on how much of the body is shown. However, the typically small size of the transmitted images on video cell phones and laptop video conferencing, or sizable pixilation of larger images that significantly blur images, considerably reduce people's ability to read emotions on the face. As technologies improve, so will the range of emotional cues perceivable by users.

Observations of infants and adults suggest that certain types of cues from mediated communication devices are likely to be more effective than others at approximating the emotional impact of physical proximity of attachment figures. Infants do not maintain interest in voices on the phone, even from close attachment figures, as anyone who has tried to talk with a baby on the telephone knows. More formal experiments with visual input show that infants need responsive, timely interactions. When confronted with the still, unresponsive face of the mother, infants became upset and eventually withdrawn after only three minutes (Tronick et al., 1978).

Adults, in contrast, can have positive emotional responses to still images. Mirror neurons and emotional responses get activated when a person sees pictures of people captured in the midst of strong emotional reactions, even when those pictured are strangers (Iacoboni et al., 2005). Research on the neurobiology of pair bonding has shown that when adults view photographs of the romantic partner, and mothers view pictures of their own children, their neuropeptide levels increase; these are the hormones associated with pair bonding—love, in other words (Young & Wang, 2004). Based on such evidence, it seems that visual cues in mediated communication are effective in helping maintain existing attachments because some of the physiological effects of attachment are triggered by seeing visuals of loved ones, with interactive visuals functioning better than still images; this warrants further investigation.

However, as neuroscientists are now discovering, mediated inputs can only approximate the physiological and psychological effects of physical proximity to others. No matter how clearly future MCTs transmit visual and voice cues, they cannot fully transmit one crucial ingredient of interpersonal communication: the resonance of minds, often called "attunement," that is a crucial aspect of the meaningful communication of emotions.

Through brain imaging and EEGs, neuroscientists are able to trace the exchange of energy within a brain and the correlating effects in the brain of a communication partner. An emotional connection happens through the mutual perception of nonverbal signals that are processed primarily by the right side of the brains of the individuals involved. As two people align their emotional states, they create a resonance of their right brain halves. This mutual alignment of emotional states is visible in the activation of similar brain structures of both communication partners. The brains tune in with each other (Siegel, 1999, 2001, 2003).

This mental attunement of minds, "feeling felt," is a pleasurable experience, difficult to express in words. Anyone who has had a beloved person, or even a pet, knows that "understanding" that can pass between beings in tune with each other, and that this feeling of connection is more powerful than words can convey. Siegel (1999) sees this experience as the key ingredient of healthy attachments: "this mutually sharing, mutually influencing set of interactions—this emotional attunement or mental state resonance—is the essence of healthy, secure attachment" (p. 117).

Because resonance between brains is probably only possible in physical proximity, the full experience of emotional communication vital to human health and social life is thus only possible through physical proximity. Mobile communication technologies might "collapse space" for communication, but can do so only asymptotically: as long as sender and receiver are not near each other, MCTs cannot completely replace the full physiological, emotional, and mental effect of physical proximity on the communication of emotions.

As far-reaching as these discoveries in the neurobiology of interpersonal interactions are, our understanding of them is still emerging. The full implications of these discoveries on human communication in general are only beginning to be discussed, while research on their implications for mediated communication is as of yet nonexistent. Because information-haves live in the hybrid physical and virtual space of the Metro/Electro Polis, further research is imperative. How do the reduced cues of interacting via mediated communication devices affect individuals physiologically, emotionally, and mentally over the long term? Interpersonal neurobiology has not yet addressed this question, but it must. Considering the ubiquity of mobile communication technologies for "keeping in touch" with friends and family, understanding how MCT use affects the neurobiology of the brain, and thus attachments, would not only benefit individuals as they come to understand the effects of MCT use, but also could significantly increase our understanding of how the human mind functions.

References

Ainsworth, M. D. S. (1972). Attachment and dependency: A comparison. In J. L. Gewirtz (Ed.), *Attachment and dependency* (pp. 97–137). Washington, D.C.: Winston.

———. (1990). Some considerations regarding theory and assessment relevant to attachments beyond infancy. In M. T. Grennberg, D. Cicchetti, & E. M. Cummings (Eds.), *Attachment in the preschool years: Theory, research, and intervention* (pp. 463–488). Chicago: University of Chicago Press.

Ainsworth, M. D. S., Blehar, M. C., Waters, E., & Wall, S. (1978). *Patterns of attachment: A psychological study of the strange situation*. Hillsdale, NJ: Erlbaum.

Baym, N. (2000). *Tune in, log on: Soaps, fandom, and online community*. Thousand Oaks, CA: Sage.

———. (2006). Interpersonal life online. In L. A. Lievrouw & S. Livingston (Eds.), *The handbook of new media* (pp. 35–54). London: Sage.

Bowlby, J. (1973). *Attachment and loss, vol. 2*. New York: Basic Books.

———. (1979). *The making and breaking of affectional bonds*. London: Tavistock.

———. (1980). *Attachment and loss, vol. 3*. New York: Basic Books.

———. (1988). *A secure base: Parent-child attachment and healthy human development*. New York: Basic Books.

Carpenter, E. M., & Kirkpatrick, L. A. (1996). Attachment style and presence of a romantic partner as moderators of psychophysiological responses to a stressful laboratory situation. *Personal Relationships*, 3, 351–367.

Collins, N. L., Guichard, A. C., Ford, M. B., & Feeney, B. C. (2004). Working models of attachment: New developments and emerging themes. In W. S. Rholes & J. A. Simpson (Eds.), *Adult attachment: Theory, research, and clinical implications* (pp. 196–239). New York: Guilford Press.

Diamond, L. M. (2001). Contributions of psychophysiology to research on adult attachment: Review and recommendations. *Personality and Social Psychology Review*, 5, 276–295.

Diamond, L. M., & Aspinwall, L. G. (2003a). Emotion regulation across the life span: An integrative perspective emphasizing self-regulation, positive affect, and dyadic processes. *Motivation and Emotion*, 27(2), 125–156.

———. (2003b). Integrating diverse developmental perspectives on emotion regulation. *Motivation and Emotion*, 27(1), 1–6.

Fox, S. A. (2000). The uses and abuses of computer-mediated communication for people with disabilities. In D. O. Braithwaite & T. L. Thompson (Eds.), *Handbook of communication and people with disabilities: Research and application* (pp. 319–336). Mahwah, NJ: Erlbaum.

Gallese, V. (2004). Intentional attunement: The mirror neuron system and its role in interpersonal relations. Virtual Workshop on Mirror Neurons. Retrieved from http://www.interdisciplines.org/mirror/papers/1.

Gallese, V., Keysers, C., & Rizzolatti, G. (2004). A unifying view of the basis of social cognition. *Trends in Cognitive Science*, 8(9), 396–403.

Goleman, D. (2006). *Social intelligence: The new science of human relationships*. New York: Bantam.

Gump, B. B., Polk, D. E., Kamarck, T. W., & Shiffman, S. M. (2001). Partner interactions are associated with reduced blood pressure in the natural environment: Ambulatory monitoring evidence from a healthy, multiethnic adult sample. *Psychosomatic Medicine*, 63, 423–433.

Hazan, C., Gur-Yaish, N., & Rholes, M. C. (2004). What does it mean to be attached? In W. S. Rholes & J. A. Simpson (Eds.), *Adult attachment: Theory, research, and clinical implications* (pp. 55–85). New York: Guilford Press.

Hofer, M. A. (1984). Relationships as regulators: A psychobiologic perspective on bereavement. *Psychosomatic Medicine*, 46, 183–197.

———. (1987). Early social relationships: A psychobiologist's view. *Child Development*, 58, 633–647.

———. (1994). Hidden regulators in attachment, separation, and loss. In M. Fox (Ed.), *The development of emotion regulation: Biological and behavioral considerations* (pp. 192–207). Monographs of the Society for Research in Child Development, 59 (2–3, Serial No. 240).

———. (2005). The psychobiology of early attachment. *Clinical Neuroscience Research*, 4(5/6), 291–300.

Houy, Y. (2005). The situationist Metro/Electro Polis: Re-imagining urban spaces in the twenty-first century. In C. Emden, C. Keen, & D. Midgley (Eds.), *Imagining the city, vol. 1: The art of urban living* (pp. 311–333). Bern: Peter Lang.

Iacoboni. M. (2005). Understanding others: Imitation, language, and empathy. In S. Hurley & N. Chater (Eds.), *Perspectives on imitation: From neuroscience to social science, vol. 1: Mechanisms of imitation and imitation in animals* (pp. 77–99). Cambridge, MA: MIT Press.

Iacoboni, M., Molnar-Szakacs, I., Gallese, V., Buccino, G., Mazziotta, J. C., & Rizzolatti, G. (2005). Grasping the intentions of others with one's own mirror neuron system. *Public Library of Science Biology*, 3(3), 529–535.

Lea, M., & Spears, R. (1995). Love at first byte? In J. Wood & S. Duck (Eds.), *Understudied relationships: Off the beaten track* (pp. 197–240). Thousand Oaks, CA: Sage.

Lewis, T., Amini, F., & Lannon, R. (2001). *A general theory of love*. New York: Vintage Books.

Matsuba, M. K. (2006). Searching for self and relationships online. *CyberPsychology and Behavior*, 9(3), 275–284.

McKenna, K. Y. A., & Bargh, J. A. (2000). Plan 9 from cyberspace: The implications of the Internet for personality and social psychology. *Personality and Social Psychology Review*, 4(1), 57–75.

Mitchell, K. J., Becker-Blease, K. A., & Finkelhor, D. (2005). Inventory of problematic Internet experiences encountered in clinical practice. *Professional Psychology: Research and Practice*, 36(5), 498–509.

Rholes, W. S., & Simpson, J. A. (2004). Attachment theory: Basic concepts and contemporary questions. In W. S. Rholes & J. A. Simpson (Eds.), *Adult attachment: Theory, research, and clinical implications* (pp. 3–14). New York: Guilford Press.

Schore, A. N. (2001). Effects of a secure attachment relationship on right brain development, affect regulation, and infant mental health. *Infant Mental Health Journal*, 22(1), 7–66.

Siegel, D. J. (1999). *The developing mind: Toward a neurobiology of interpersonal experience*. New York: Guilford Press.

———. (2001). Toward an interpersonal neurobiology of the developing mind: Attachment relationships, "mindsight," and neural integration. *Infant Mental Health Journal*, 22(1), 67–94.

———. (2003). An interpersonal neurobiology of psychotherapy: The developing mind and the resolution of trauma. In M. F. Solomon & D. J. Siegel (Eds.), *Healing trauma: Attachment, mind, body and brain* (pp. 1–56). New York: W. W. Norton.

Solomon, M. F., & Siegel, D. J. (Eds.). (2003). *Healing trauma: Attachment, mind, body and brain*. New York: W. W. Norton.

Tronick, E., Als, H. L., Adamson, S., Wise, S., & Brazelton, T. B. (1978). The infant's response to entrapment between contradictory messages in face-to-face interaction. *Journal of Child Psychiatry*, 17, 1–13.

Ward, C. C., & Tracey, T. J. G. (2004). Relation of shyness with aspects of online relationship involvement. *Journal of Social and Personal Relationships*, 21(5), 611–623.

Wellman, B., & Gulia, M. (1999). Virtual communities as communities: Net surfers don't ride alone. In M. Smith & P. Kollock (Eds.), *Communities in cyberspace* (pp. 167–194). New York: Routledge.

Young, L. J., & Wang, Z. (2004). The neurobiology of pair bonding. *Nature Neuroscience*, 7(10), 1048–1054.

·5·

DISPLACING PLACE WITH OBSOLETE INFORMATION AND COMMUNICATION TECHNOLOGIES

Julie Newman

My friend Jim's eleven-year-old daughter Katie recently requested permission to have her own cell phone (in case of an emergency and to text message) and computer (for email and access to the World Wide Web for research papers). Jim was already planning to buy Katie an iPod for her twelfth birthday. In the span of Katie's life, based on today's standard life cycle (production to disposal) of cell phones and computers, she will own and dispose of approximately thirty-five cell phones (a new phone approximately every two years) and twenty-three computers (a new computer approximately every three years). Our ever-increasing demand for accessible, affordable, smaller, faster, and more fashionable information and communication technologies (ICTs) is leaving states and municipalities with the challenge of what to do with these devices when they are no longer considered to be of value to users.

ICTs are influencing the manner in which we conduct business, seek information, maintain social relationships, entertain ourselves, and more. The issues at hand are the environmental and human health risks associated with the production and disposal phases of rapidly evolving ICTs. The framework for capturing the challenge of balancing technological expansion with the impact on environmental and human health can be referred to as sustainability. Sustainability is grounded in the assumption that the pursuit of economic viability as an end in itself will not necessarily maintain or enhance ecological health and human well-being.

In a nonsustainable society, people tend to disconnect their desire for and acquisition of products from the life-cycle impacts of products. This is not a new behavioral phenomenon in our consumer-driven society. What has changed is that our usual means of disposal—placing our discards in the garbage can or on the curb—are no longer acceptable for disposing of used electronic devices, such as computers, cell phones, and other ICTs.

Rapid advances in ICTs have led to an expanding category of waste now referred to as electronic waste (or e-waste). This is a growing waste stream predominantly in countries affiliated with the Organisation for Economic Co-operation and Development (OECD), which have highly saturated high-tech markets (Widmer et al., 2005). E-waste is composed of obsolete electronic equipment. Estimates of e-waste range from 1 to 8 percent of the municipal waste stream in the United States. However, this stream is predicted to increase at three times the rate of municipal solid waste in the next decade (Royte, 2005). That may not appear to be much, but the ratio of hazardous materials within the standard e-waste stream is greater than in the average solid waste stream (Widmer et al., 2005). The e-waste stream is projected to increase before leveling out, as the market for personal computers and other electronic devices in OECD countries has yet to be saturated (Environmental Protection Agency [EPA], 2005).

Personal computers make up only a portion of the e-waste stream; other emerging and growing e-waste contributors include personal digital assistants (PDAs), digital media players, cell phones, computer games, and related peripherals. By 2010, an estimated 250 million personal computers will become obsolete in the United States, contributing to the projected 400 million units of expected obsolete electronic equipment (Sells, 2005). This number is equivalent to discarding 68,000 computers per day for a year. Added to that will be an estimated 650 million cell phones, discarded at a rate of 130 million per year, according to the EPA (2002). Moreover, the EPA estimates that, appallingly, only one out of ten obsolete computers are properly disposed of and recycled.

Recyclers are processing more than 1.5 billion pounds of electronic equipment annually (Schneiderman, 2004). As new models of computers, cell phones, digital media players, and other such devices are shrinking in size, the replacement cycle is accelerating, because people want the newest, sleekest models, leading, ironically, to more e-waste. In 2003, the International

Association of Electronics Recyclers (IAER) determined that 900 million pounds of usable materials were captured from the 1.5 billion tons of waste produced (IAER, n.d.). A Carnegie Mellon University study concluded that a total of 170 million cubic feet of landfill space would be needed to dispose of 55 million whole personal computers plus the 10 percent of scrap recycled computers, or roughly one acre piled four thousand feet high (Matthews et al., 2003).

Uncertainty about the long-term negative impacts of improper disposal has been the impetus for municipal, state, federal, and international (European Union [EU], Canada, and Japan) regulations. Seventy percent of the heavy metals that are found within U.S. landfills today, such as cadmium, mercury, and lead, are from computers. The materials in computer monitors have always been categorized as hazardous waste (EPA, 2005). In 1995, cathode ray tubes (CRTs) containing lead were officially designated as hazardous waste in the United States and officially banned from disposal in the solid waste stream. There are more than a thousand different substances found in e-waste, many of which are toxic. Some areas of concern include:

- Uncertainty about the long-term human health impacts on workers and nearby communities as a result of long-term exposure to emissions and other manufacturing processes;
- Uncertainty about the long-term human health effects of landfill leaching; and
- Energy required to manufacture and operate computers multiplied by the growing demand (Sells, 2005).

In addition to the human and environmental health concerns, another essential issue related to sustainability is the excessive waste of valuable nonrenewable materials. Electronic devices are created from many valuable resources, including metals, such as gold and copper, as well as engineered plastics and glass (EPA, 2002). Whether mined or manufactured, these materials require energy to extract or process. Many parts of obsolete electronics can be profitably refurbished and reused. As we discard electronic devices, we are also discarding valuable nonrenewable resources.

The U.S. Government Accountability Office estimates that, of the electronics that are captured for recycling in the United States, 50 to 80 percent are shipped to Asia or Africa (Hileman, 2006; Silicon Valley Toxics Coalition, 2002). It is this 50 to 80 percent that has led to increased regulations regarding international trafficking of electronic waste. In 1989, the Basal Convention was initiated to respond to the existing and anticipated increase of hazardous waste export to non-OECD countries. The adoption and implementation of the Basal Convention recommendations ensure proper transboundary trade and disposal (Basal Action Network, n.d.). In 1992, the Basal Convention was endorsed by both OECD and non-OECD countries in an effort to limit and ultimately end the export and improper disposal of hazardous materials to countries within Africa and Asia, and developing nations elsewhere. Unfortunately, there is little or no financial incentive for individual households to recycle, which means that many people either stockpile or improperly dispose of their obsolete electronics. Moreover, materials recovered through electronic recycling tend to be worth less than the combined costs of collecting and disassembling—to separate usable materials from discards—the obsolete devices.

Here is what happens to electronic products that are taken back by manufacturers or routed to recycling businesses:

- The glass CRT is either returned to a manufacturer and made into a new CRT, or sent to a smelter, where the lead is recovered and recycled;
- The plastic housing is ground into smaller pieces and recycled for use in various items, such as highway dividers and pothole mix;
- Circuit boards, chips, and other parts are reused to repair or upgrade older electronics or recycled for their scrap value;
- Metal components are separated and sold for their scrap value (EPA, 2002).

Waste disposal for all products and materials is an essential issue facing our consumer society, from households and municipalities to universities and large businesses. The process of waste collection and disposal can be a major burden on taxpayers, manufacturers, collectors, and governments. In the United States alone, solid waste disposal costs taxpayers and local governments upwards of $40 billion a year. The hidden costs of our waste management system are buried in the standard collection procedure that enables us to leave our waste at the curbside (household) or deskside (office) for pickup. This system has prolonged and encouraged us to be a "throw-away" society: we are fooled into thinking that there is an "away" and that when our goods leave the corner they are being disposed of safely. Shockingly, 80 percent of what we consume in the United States is discarded after a single use. ICTs and other electronic devices classified as e-waste require a new level of consumer and producer engagement to ensure their proper disposal, so that we can foster sustainable, ecologically healthy communities.

E-Waste Regulations in the United States: An Evolving State of Classifications and Displaced Responsibility

The fast-growing e-waste stream in the United States has led to a series of municipal, state, and federal regulations and to the proliferation of waste disposal and recycling businesses. In 1993, the EPA updated the existing Resource Conservation and Recovery Act (RCRA), proposing to streamline the disposal of certain hazardous materials. In 1995, items such as batteries, pesticides, and thermostats were designated as universal waste. These regulations, which primarily apply to large businesses, also impact average households. By 2005, the universal waste category was expanded to include lamps as well as items containing mercury, such as thermometers, computers, and cell phones (EPA, 2005). Items such as computers, PDAs, and cell phones are designated as hazardous material only when the object is considered to be an end-of-life discard. Regulations governing the collection and management of these widely generated wastes facilitate environmentally sound collection and proper recycling or treatment.

Today, large businesses and institutions, such as universities, can no longer ship their obsolete electronics to landfills. While businesses are prohibited from disposing of electronics in landfills or incinerators, average household consumers are charged with determining whether their state or municipality regulates household electronic materials, because e-waste disposal regulations vary from state to state.

Universal waste regulations have two other primary intentions: to ease the regulatory burden on retail stores and others that wish to collect and properly dispose of these wastes, and to encourage the development of municipal and commercial programs to reduce the quantity of these wastes going to municipal solid waste landfills or incinerators. Universal waste regulations also require that wastes go to appropriate treatment or recycling facilities.

The growing quantity of e-waste has led to emerging responses from individual states. For example, California, Maine, and Massachusetts have passed regulatory laws banning electronic equipment from landfills and incinerators; another twenty-six states are considering similar legislation. In California, as of January 2005, an advance recovery fee of six to ten dollars is collected on all new televisions and computer monitors sold in the state (Hileman, 2006). The fee, which varies based on the size of the device, is used to reimburse nonprofit and for-profit collectors and recycling businesses. California recyclers are not allowed to send obsolete devices overseas unless they can prove that the waste will be handled in compliance with the standards developed by the OECD. In 2005, California collected an estimated $65 million from the program, covering the cost of proper disposal for 50 million pounds of electronics for that year.

In Maine, legislators passed a law in 2004 similar to that in the EU that places the responsibility on manufacturers. Computer and television producers that sell in Maine must finance the take-back and recycling of the discarded electronics. The consumer is responsible for transporting the material to a designated drop-off point, and the producer is responsible for final collection and recycling (Hileman, 2006).

Similar to Maine and California, in 2000 Massachusetts legislators banned CRTs from landfills and incinerators. The Massachusetts Department of Environmental Protection program encourages residents to determine if their municipality has a computer and electronics recycling program, bring used items to local retail stores or businesses that recycle or reuse discarded equipment, or donate electronic devices to established programs.

The U.S. House and Senate have both held hearings on electronic waste, and many bills have been introduced into Congress attempting to provide mechanisms for capturing the growing stream of e-waste to ensure proper demanufacturing and disposal. This is not the first time that mechanisms to increase the capture rate of end-of-life electronic items have been discussed at the federal level. In 1997, a provision of the Taxpayer Relief Act entitled the 21st Century Classrooms Act included a clause that enabled private companies to donate old computers and related equipment to schools (Royte, 2005). Businesses that donate their used computers to public or private K–12 schools are able to take deductions on their tax returns for the full purchase cost of the items. A clause in the act prevents businesses from dumping outdated equipment; companies must document that the equipment is two years old or newer.

In 2005, the Electronic Waste Recycling Promotion and Consumer Protection Act (S-510) was introduced in the Senate. This act assumes that:

- Discarded electronic waste tends to be handled as municipal solid waste in communities and states where legislation has not been enacted;
- Discarded electronic waste accounts for 40 percent of the lead and 70 percent of the heavy metals found in landfills that, when handled improperly, can be released into the environment, leading to contamination of air and groundwater and posing a significant threat to human

health, including potential damage to kidney, brain, and nervous system function, and cancer in cases of excessive exposure;

- Properly recycling electronic waste can conserve resources and minimize the potentially harmful human and environmental health impacts of those materials found in computers, televisions, and similar electronic products.

The goal of the legislation is to facilitate people's access to recycling services, and to improve the efficiency and use of e-waste recycling ("Electronic Waste Recycling," 2005). The intention is to increase incentives for computer capture and disposal and ban certain electronic items from landfills. For example, businesses that collect discarded electronics would receive a tax credit of $8 per unit ("Electronic Waste Recycling," 2005). Other provisions require that all federally owned computers and monitors be recycled and mandate an EPA study to assess the feasibility of a national e-waste recycling program.

While there are many budding efforts in the United States, particularly at the state level, to begin to cope with this growing waste stream, such efforts are not as aggressive as the recently adopted Waste Electrical and Electronic Equipment (WEEE) directive being implemented by countries in the EU.

Regulations in the European Union: Placing Responsibility

A study published in 2004 by the United Nations University in Tokyo analyzed the life cycle of computers from development to disposal and called upon the Japanese government "to extend the life of personal computers and slow the growth of high-tech trash" (Williams, 2004, p. 1). The study concluded that the manufacturing process of the average desktop computer requires ten times its weight in both fossil fuels and chemicals. While the EU has not yet called for an extended life cycle, it has developed a take-back mechanism that provides an impetus for improved manufacturing processes by placing responsibility for disposal on producers.

In 2002, the EU developed the WEEE directive that requires the retailer and producer or contracted service to take back electronic products once the items are deemed obsolete (Lifset & Lindquist, 2003). Using a life-cycle analysis framework, electronics designers are working to develop processes that eliminate lead-based products, such as solder, as well as other banned substances. Hand in hand with the WEEE directive is the Restriction of Hazardous Substances (RoHS) directive, which aims to reduce the type and amount of hazardous substances that could leach out of landfills and into nearby water sources or groundwater, as well as to prevent the release of toxic substances from unintended incineration. RoHS guides the design and development of the electrical devices that tend to contain heavy metals and toxic substances (Hoske, 2004). If RoHS is implemented successfully, strict maximums for substance content will be established, which will reduce the amount of hazardous substances in ICTs.

A new approach for solving the imminent e-waste problem, based on a life-cycle framework, is manifesting in Extended Producer Responsibility (EPR) regulations. One commonly used definition of EPR, as adopted by the OECD, suggests that the responsibility of the producer extends to the postconsumer stage of the product's life cycle, including disposal (Widmer et al., 2005). In essence, EPR places a framework of continued accountability on producers over the life cycle of their products. The EU has taken the lead in adopting EPR principles, as is illustrated in the recently adopted WEEE standards. The EPR framework was initiated in Germany following a projected shortage of landfill space. This led to the adoption of a policy that required producers to develop recyclable and nonhazardous packaging materials.

EPR mechanisms are being adapted and institutionalized throughout much of the industrialized world, with the unfortunate exception of the United States. Many European companies, including car manufacturers, are now making their products easier to disassemble and recycle at end of the life cycle. The evolution in thinking about how to minimize the life-cycle impact of the product is captured in this explanation:

> Transferring responsibility for waste to the producer is one of the best ways to force changes "upstream." When companies have to worry about what's going to happen to their product when it becomes waste it affects all of their decisions about product design and material use. When companies reduce the amount of material they're using to package a product, that means less raw material has to be extracted. You avoid all of the major environmental impacts of extraction, material processing, and manufacturing. (Fishbein, 1998, p. 1)

As the WEEE directive in the EU begins to take shape, the question of individual and institutional responsibility still lingers. While there is a market-driven incentive to create new and innovative ideas for smaller, faster, and sassier new ICTs, how can we be sure to apply the same level of creativity to reducing our e-waste stream?

Placing Discards in the Proper Places

Although there are no federally mandated regulations regarding discards that manufacturers must respond to in the United States, a number of multinational companies have developed policies and practices, many in response to the WEEE directive. However, effectively communicating to consumers the most appropriate steps for properly disposing of ICTs has not yet become standard. Moreover, in light of the ever-increasing market for ICTs, such as cell phones and laptops, manufacturers are continually updating their disposal policies, because there are competitive advantages associated with regulated collection and disposal. Examples of this evolving market include policies from Verizon Wireless, T-Mobile, and Cingular. These companies, to varying degrees, have all integrated language and data about their collection strategy on their Web sites. Verizon Wireless recognizes the importance of recycling wireless devices and accessories to help ensure a cleaner environment for future generations, and the company has several initiatives for promoting the environmentally safe reuse or disposal of these products. According to their corporate Web site (http://www22.verizon.com), Verizon Wireless has collected more than 200 tons of electronics waste and 150,000 pounds of batteries, which have been kept out of landfills. In 2005 alone, 275,000 phones were recycled in an environmentally safe way through their HopeLine program.

Computer companies have been developing and promoting computer disposal regulations and mechanisms since the 1990s, and product take-back programs are now available through a variety of manufacturers. Apple, Dell, Hewlett Packard, IBM, and others provide institutional computer take-back programs for a cost, ranging from nothing to $45 per unit (computer processing unit [CPU], monitor, and so on). Computers deemed to have value are sold, enabling the computer company to recuperate some of the disposal cost. For example, Apple offers U.S. customers the option of recycling their unwanted personal computers, regardless of the manufacturer. All equipment is recycled domestically, and no hazardous material is shipped overseas. Dell offers consumers convenient, free recycling for Dell end-of-life products. According to Dell's Web site (http://www.dell.com), the company is taking responsibility for continually improving the environmental design aspects of its products and end-of-life management. Dell encourages this same level of responsibility from other producers throughout the electronics industry.

Corporate policies and procedures continue to evolve with the industry and now reflect the growing concerns about the environmental and human health impacts of improperly disposed obsolete devices coupled with greater awareness of waste regulations in the United States and the WEEE directive in the EU.

Many business opportunities and corporate partnerships have grown out of discussions about how to better manage and decrease the e-waste stream. One such partnership in the United States is the Federal Electronics Challenge, which encourages federal facilities and agencies to purchase greener electronic products, reduce the impact of electronics during use, and properly manage obsolete electronics. This partnership grew out of an analysis of the federal government's use of electronics that demonstrated that the federal government purchases more than $60 billion of electronic equipment annually. At this rate, the government has the opportunity to take a leadership role in tackling this complex issue. On the industry front, a similar organization, IAER, represents industry interests for developing effective mechanisms for managing the life cycle of electronic products.

Innovative Manufacturing Mechanisms and Disposal Options

When people decide to buy a new cell phone, laptop, or PDA, what in the system triggers them to think about product disposal two to four years down the road? A systemic approach is necessary for creating a sustainable society. We have to replace responsibility on manufacturers, municipalities, institutions, and consumers.

Manufacturers have the opportunity to develop and implement mechanisms to reduce environmental and human health impacts of production and disposal processes. Greener ICTs should incorporate fewer toxic materials, recycled materials, energy efficiency, and minimal packaging on the production front. On the sales end, products should have leasing or take-back options, and they should display ecolabeling and certification.

In addition to manufacturers, institutions, corporations, and municipalities are beginning to develop sustainable waste management systems that minimize the impact of disposal on ecosystem health and human health through source reduction, reuse, recycling, treatment, and landfill disposal. However, until waste production is recognized as a cultural component of a system, the current trend of consumption will continue to displace the center of responsibility from the producer to the disposer.

This chapter began by stating that there is a tendency for people in a non-sustainable society to disconnect their desire for and acquisition of products from awareness of the life-cycle impact of the products. Whose responsibility is it to ensure that the impact of e-waste disposal is minimized and that our communities do not become contaminated by the improper disposal of electronic equipment? How can ICTs be used to communicate and encourage behavior that leads to informed consumer choices and proper disposal of our obsolete electronics? These are questions we must address. Fundamentally, we must recognize that although we are accustomed to throwing things away, we cannot continue to deceive ourselves into thinking that there is no "away." In a sustainable society, we can certainly integrate ICTs into our daily lives, but even as we displace place with our increasing capabilities for communicating at any time from anywhere, we need to develop a consumer model that favors environmental and human health in all places.

References

Basal Action Network. (N.d.). The Basal Convention. Retrieved October 1, 2006, from http://www.ban.org.

Electronic Waste Recycling Promotion and Consumer Protection Act of 2005. (2005). S. 510, 109th Cong., 1st Sess.

Environmental Protection Agency (EPA). (2002). Municipal solid waste in the United States: 2000 facts and figures. EPA Publication No. EPA530-R-02–001. Washington, D.C.: U.S. Government Printing Office.

———. (2005). Waste wise update. EPA Publication No. EPA530-N-00–007. Washington, D.C.: U.S. Government Printing Office.

Fishbein, B. (1998). Extended producer responsibility. *Environmental Manager: Environmental Solutions That Make Good Business Sense*, 10(1), 1–12.

Hileman, B. (2006). States strive to solve burgeoning disposal problem as more waste ends up in developing countries. *Chemical and Engineering News*, 2(84), 18–21.

Hoske, M. T. (2004). Life cycle environmentalism. *Control Engineering*, 51(1), 36–38.

International Association of Electronics Recyclers. (N.d.). IAER electronics recycling industry report. Retrieved September 1, 2006, from http://www.iaer.org/.

Lifset, R., & Lindquist, T. (2003). Can we take the concept of individual producer responsibility from theory to practice? *Journal of Industrial Ecology*, 7(2), 3–6.

Matthews, S., McMichael, F., Hendrickson, C., & Hart, D. (2003). Disposition and end-of-life options for personal computers. Green Design Initiative Technical Report. Pittsburgh, PA: Carnegie Mellon University.

Organisation for Economic Co-operation and Development (OECD). (2001). *Extended producer responsibility: A guidance manual for governments*. Paris: OECD.

Royte, E. (2005). *Garbageland: On the secret trail of trash*. New York: Time Warner Book Group.

Schneiderman, R. (2004). Electronic waste: Be part of the solution. *Electronic Design*, 52, 47–54.

Sells, B. (2005). Between the lines. *Waste Age*, 36(4), 120–123.

Silicon Valley Toxics Coalition. (2002). *Exporting harm: The high-tech trashing of Asia*. Seattle, WA: Silicon Valley Toxics Coalition, Basal Action Network.

Widmer, R., Oswald-Krapf, H., Sinha-Khetriwal, D., Schnellmann, M., & Böni, H. W. (2005). Global perspectives on e-waste. *Environmental Impact Assessment Review*, 25, 436–458.

Williams, E. (2004). International activities on e-waste and guidelines for future work. Proceedings of the Third Workshop on Material Cycles and Waste Management in Asia NIES (pp. 1–11). Tsukuba, Japan.

· 2 ·

MOBILE INNOVATIONS

· 6 ·

CYBER-CRIME ON THE MOVE

Matthew Williams

Deviant behavior that manifests on or is facilitated by networks of computers is a well-documented phenomenon. Theoretical, empirical, and technical aspects of what has been dubbed "cyber-crime" have been the focus of study for researchers within a variety of disciplines (see Wall, 2001; Williams, 2006). Of the claims made across writings, one overarching commonality remains constant: the Internet and its constituent technologies distantiate time and space (Giddens, 1990). This process facilitates criminal enterprise while also serving to complicate its control. Over the past decade, industry victimization surveys show how cyber-criminals utilize networks of computers to commit multiple frauds in compressed periods of time over vast distances (Department of Trade and Industry, 2006; Richardson, 2003). The same surveys illustrate how private companies, governments, and domestic Internet users are ill prepared for the array of threats they face online. Further, the ability of cyber-criminals to transcend traditional geographic boundaries presents criminal justice systems across the globe with their greatest challenge. Most recently this challenge has been exacerbated by the illicit use of mobile technologies, such as mobile (cell) phones and Wi-Fi. (Wi-Fi, or "wireless fidelity," refers to local area networks that support simultaneous wireless broadband Internet connections. See Jassem, chapter 2 in this volume.) In particular, the insecurity of wireless networks is providing cyber-criminals with an array of opportunities to steal proprietary information from those on the move (Chalmers & Almeroth, 2004). In relation to the former, surveys are beginning to show how digital harassers are adding mobile phones to their arsenal (NCH, 2005). The increasing ubiquity of mobile technology and our dependency on communication at a distance leave us exposed to the actions of sophisticated hackers, online fraudsters, and abusers. This chapter examines the characteristics of various types of cyber-crime and explores the extent to which they have migrated to mobile communication platforms.

Temporal and Spatial Dimensions of Cyber-Crime

Crime, like any other social practice, is space and time dependent. Crimes that are committed within the "terrestrial" domain that make no use of information technology can be said to be significantly restricted by place and time. Terrestrial place and space afford and restrict criminals in multifaceted ways: at a basic level, urban and rural spaces shape the form, severity, and prevalence of crimes. Research consistently shows that domestic burglaries, vehicle theft, and violent offenses are far more common in urban spaces compared with rural spaces (Aust & Simmons, 2002). Both rural and urban spaces promote and shape specific population demographics, the built environment, and relations between people, which in turn either provide opportunities for crime (either in terms of environmental, psychological, or social opportunities) or serve to prevent it (for example, via social or psychological control or reduced environmental opportunities). While terrestrial crimes are spatially sensitive and dependent, they are also geographically bounded. With few exceptions, crimes require that perpetrator and victim (or their property) meet in time and space. While physical convergence is necessary for a terrestrial crime to take place, the event must also be recognized as criminal by the victim (or bystander). Only then can an act and the perpetrator be labeled as criminal (Becker, 1963).

Unlike terrestrial crimes, cyber-crimes committed via static and mobile technologies displace both time and space. Cyber-crimes can be said to be temporally and spatially fluid when illicit actions in one temporal-spatial boundary have consequences outside of that restriction. Harassers, fraudsters, drug dealers, and human traffickers, to name a few, can utilize information technology to organize criminal tasks and victimize other technology users over unprecedented distances in compressed periods of time. In addition, the recognition and detection of deviant or criminal behavior committed over computer networks is complicated by the technologies' displacing capacity: what is considered criminal in the locale of the victim may not be in the locale of the perpetrator. Such jurisdictional inconsistencies make the detection and prosecution of cyber-criminals less likely than that of their terrestrial counterparts. The increasing use of mobile technologies to communicate and wirelessly connect to the Internet further expands cyber-criminal opportunity, inviting previously ill-equipped would-be deviants into criminal behaviors.

Cyber-Crime and Mobile ICTs

Before categorizing types of cyber-crime, it is important to distinguish what is criminal behavior and what is not in relation to static and mobile technology use or misuse. As advances in technology escalate at a rate unparalleled by everyday institutional mechanisms, it is no surprise that the law is slow to respond to high-tech criminal activity. Where advances have been made, it is questionable whether the scope of new and adapted bodies of jurisprudence can remain ahead of deviant enterprise, especially in the case of new mobile technology misuse. While certain forms of hacking, obscene electronic materials, and online stalking have been met with legal rationalization, many other forms of computer-related activities escape regulation due to their esoteric nature. For example, certain activities associated with mobile phones, such as text bullying (NCH, 2005), escape legal regulation but are still arguably harmful. As a result, it has become acceptable, and more analytically fruitful, to consider harms, instead of crimes, in relation to technology use or misuse. Examining how one's behavior on a computer network can negatively affect an individual, be it financially or psychologically, allows for a greater scope in understanding both criminal and noncriminal computer-related activities.

While legal systems struggle to tackle the definitional problem, academic writers have attempted to delineate cyber-crimes in several ways. David Wall (1998) attempts to apply a classification based upon technological embeddedness. First, there are low-end users who utilize the technology to facilitate terrestrial criminal activity. For example, both the Internet and mobile phones have been used to organize violent exchanges between football hooligans in the UK (British Broadcasting Corporation, 1999). Similarly, online pedophiles are increasingly utilizing both the mobile Internet and phones to groom children (Gardner, 2003). Second, Wall (1998) identifies medium technology users who have created new opportunities for harmful activity that are recognized by bodies of law. Examples include the creation of new kinds of obscenity though computer-generated images (pseudo-photographs), certain forms of computer scams, such as phishing and pharming (Ponemon Institute, 2004), and mobile phone scams (McCormack, 1996; O'Malley, 1995). Last, high-end technology users are engineering entirely novel forms of deviant activity that escape legal rationalization. Many of these forms of deviance manifest and wreak their consequences within online environments. Forms of avatar abuse within virtual-reality graphical environments provide a good example (Williams, 2006). One of the most extreme documented examples appears in a paper by Richard MacKinnon (1997), who analyzes a case of virtual rape that occurred in the online community LambdaMOO. As the mobile Internet becomes more sophisticated and user-friendly, it is likely that these forms of deviance will migrate to portable platforms, increasing their prevalence and impact. The following sections turn to specific types of cyber-crime/deviance that currently manifest both on static and mobile forms of information and communication technologies (ICTs).

Hacking

Hacking Web sites and other computer systems is considered cyber-trespass. Three categories of illicit activity are conducted by hackers. The first type of activity is the deliberate manipulation of data, such as Web pages, so that they misrepresent the organization or person they are supposed to represent. The Web sites of several political parties have been targeted by hackers on the run-up to general elections in the UK, and manifestos have been rewritten in sarcastic and satiric fashion. In particular, hacktivists attempt to draw attention to their ideological position by defacing Web sites, such as those representing commercial or governmental interests. Notable attacks upon UK government Web sites occurred in August 2000, when a well-known hacktivist going under the name "Herbless" was successful at defacing over ten government Web sites with an aim to criticize official policy on smoking. Herbless had already achieved some notoriety, having hacked the UK Cabinet Office Web site a month previously.

Terrestrial forms of vandalism are material crimes because they have a physical presence. Conversely, acts of online vandalism have no tangible element, making them immaterial. Although the defacement inflicted by a hacker on a Web site is visual, there is actually no physical damage—and repairing the damage done usually involves nothing more than downloading the original file of computer code to replace the corrupted one. The effects of online vandalism, however, are said to be disproportionate, because the damage to corporate, political, and personal reputation can be substantial.

The remaining two kinds of activity associated with hackers are cyber-espionage and cyber-terrorism. E-spies break access codes and passwords to enter classified areas on computer networks. The primary aim of the e-spy is to appropriate classified knowledge. In comparison, e-terrorism can take many forms, including denial-of-service attacks in which entire servers are brought to a standstill, halting business and sometimes even whole economies. Richard Clarke, former White House terrorism czar, highlighted the risks associated with new networked technologies:

> [CEOs of big corporations] think I'm talking about a 14-year-old hacking into their Web sites. I'm talking about people shutting down a city's electricity, shutting down 911 systems, shutting down telephone networks and transportation systems. You black out a city, people die. Black out lots of cities, lots of people die. It's as bad as being attacked by bombs. . . . Imagine a few years from now: A president goes forth and orders troops to move. The lights go out, the phones don't ring, the trains don't move. That's what we mean by an electronic Pearl Harbor. (Quoted in Sussmann, 1999, pp. 452–453)

While a little emphatic, Clarke's concerns were partially justified. It is no secret that military strategists are preparing to counter "information warfare," so defined when intruders enter major computer systems and cause damage to their contents, thus causing considerable damage to the target. It is known that such intruders can infiltrate and tamper with national insurance numbers (the UK equivalent of U.S. social security numbers) and tax codes, bringing economies to a standstill (Sussmann, 1999).

It is no surprise that the activities of hackers have migrated to mobile platforms (see Chalmers & Almeroth, 2004). The burgeoning of freely accessible wireless networks in hotels, airports, cafés, and many other public spaces encourages mobile access to the Internet that is rarely secure. Increasingly sophisticated and easily accessible programs known as "packet sniffers" are being utilized by hackers to capture credit card numbers, passwords, and bank account details as they are transmitted via wireless networks to their destination servers. These details are then used to facilitate online frauds. However, the threat is variable, and it seems that domestic users are at more risk than business users. Many companies now require their employees to use virtual private network (VPN) software when connecting to the company server via wireless networks. This provides a level of secure encryption of sensitive information that domestic users can rarely afford.

Accessing the Internet via mobile phones is still relatively uncommon. A recent survey of fifteen hundred mobile phone users in the UK showed that 73 percent reported not using Internet services via their phone. Slow-loading pages and navigation problems were reported as the main barriers to mobile Internet access (Telecommunications, Network and Security Research Group [TNS], 2006). However, such low Internet take-up does not prevent hackers from targeting mobile phone users. While traditional Internet hacks may be unsuitable, key insecurities in Bluetooth technology are leaving mobile users open to attack. Freely available software such as "bluesnarf" enables hackers to download a mobile phone's entire contents, including phone numbers, calendar information, images, and video. What is evident from such trends is that a gap seems to be opening between society's increasing dependence upon mobile communication technologies and its ability to maintain and control their abuse and misuse.

Malware

Malware are computer programs that are designed to disrupt normal computational operation, including the deletion of data and the illicit transmission of personal details. Viruses, a form of malware, can be understood as malignant computer code that self-replicates and spreads by inserting copies of itself into other executable code or documents on a computer, network, or software package. The infected hardware and software become "hosts" to the computer virus. For viruses to spread, they have to attach to known computer codes on the host (such as common code within operating systems such as Windows XP). Viruses can be written to perform a vast array of functions, from disrupting everyday computer software to destroying or manipulating proprietary data, resulting in financial losses. They typically take up computer memory used by legitimate programs. As a result, they often cause erratic behavior and can result in system crashes.

Other types of malware include worms and Trojan horses. Worms, unlike viruses, can self-propagate without attaching to a known code on a host computer. They usually take advantage of computer and network file-transmission capabilities and as a result harm business networks and stifle bandwidth. In extreme cases, worms are used by cyber-criminals to subvert whole networks of computers, bringing them under their illicit control. Often these networks, known as "bot-nets," are used to send out spam email over the entire Internet. Cyber-criminals have also been known to hold companies to ransom using worms and associated bot-nets to threaten denial-of-service attacks. Cryptoviral extortion, where proprietary information is encrypted and companies are instructed to hand over funds before the cyber-criminal decrypts the data, is also common. The most damaging worm to date, Mydoom, almost brought the Internet to a standstill in 2004. The worm was responsible for creating bot-nets and sending spam to targeted companies in distributed denial-of-service attempts.

Trojan horses are malicious computer programs that masquerade as something harmless or interesting to the user. They are often distributed via email. Once executed, the program can perform an array of illicit functions, including sending viruses, creating bot-nets, and uploading or downloading files. Most significant is the ability of Trojan horses to install spyware on host computers and networks. Such spyware is capable of logging users' keystrokes and taking screen shots in an attempt to access sensitive information and data. Users' bank details, passwords, and security information can easily be recorded and sent back to the cyber-criminal. Not surprisingly, malware infection is the most common type of business and personal computer security breach. Security measures are least effective against malware infection, and incidents are on the increase. For businesses with few security measures in place, malware poses a significant threat to financial security. Globally, annual costs from malware infection are in excess of $14 billion, with the most notorious virus, the Love Bug, costing businesses over $8.5 billion (Computer Economics, 2005).

Until recently, malware has been able to infect only personal computers. In 2004, the first mobile phone virus, Cabir, was created. It has been followed by others that are increasingly sophisticated and malicious. To date, only the most advanced handhelds, smartphones, are susceptible to attack. These phones are designed to function much like personal digital assistants (PDAs) and have the capacity to run mobile versions of Microsoft's Windows. Security and phone companies are currently complacent regarding the threat of mobile viruses, as only certain technologies can be targeted. What is evident from previous trends is that as mobile phones become more sophisticated so will virus writers' efforts to subvert the technology.

Cyber-Fraud

Financial fraud is an umbrella term for a myriad of illicit activities. Computer networks and the Internet have facilitated financial fraud and have allowed previously ill-equipped would-be criminals to illicitly appropriate funds. There are currently three common methods that facilitate this type of cyber-crime. The first, Trojan horses, have already been discussed. Briefly, these malware programs surreptitiously record computer activity via key loggers and screen grabbers to identify and communicate back to the cyber-criminal sensitive security information. The other two methods are phishing and, more recently, pharming. Phishing attacks use both social engineering and technical subterfuge to steal consumers' personal identity data and financial account credentials. Such schemes use "spoofed" emails to lead consumers to counterfeit Web sites designed to trick recipients into divulging financial data such as credit card numbers, account usernames, and passwords. By hijacking brand names of banks, e-retailers, and credit card companies, these e-criminals often convince recipients to respond. Pharming removes the social engineering element with a technological fix. Malware is typically installed on a computer via a Trojan horse that automatically directs the user to the spoof Web site address when the original bookmark is selected.

In 2004, overall annual costs to banks as a result of phishing varied between $500 million (Ponemon Institute, 2004) and $1.2 billion (Gartner Research, 2004) in the United States. In the UK, bank losses as a result of phishing equalled £12 million in 2004 (Association for Payment Clearing Services [APACS], 2004). These amounts are significantly less than those resulting from "traditional" credit card fraud. However, appropriating banking information from businesses is only one of the consequences of this cyber-crime. Business Web sites have also been targeted in phishing and pharming scams and as a result have lost customers and reputations. While difficult to quantify, arguably these costs would greatly outweigh those previously cited.

Wall (2001) identifies one further type of cyber-theft that is becoming more prevalent: the appropriation of intellectual property when, for example, music or video has been recorded and digitally reproduced and distributed over computer networks. The most notorious successful prosecution for this type of cyber-theft was in the case of *A&M Records, Inc. v. Napster, Inc.* (2000), in which the defendant was accused of distributing and selling copyrighted music. The Digital Millennium Copyright Act of 1998 (DMCA) was introduced to update U.S. law for the digital age. Certain aspects of the act made Internet service providers (ISPs) liable for copyright violations. While the DMCA was comprehensive enough to satisfy the World Intellectual Property Organization's (WIPO) treaties, there were still concerns over its inadequate protections for copyright owners. The proposed Berman Bill of 2002 in the United States allows copyright owners to effectively violate the law in protection of their products. The bill allows for the hacking of any computer that is downloading copyrighted material from a peer-to-peer network. The bill allows for action tantamount to vigilante justice, with copyright owners acting as prosecutor, judge, and jury. Due to opposition from digital civil libertarians, it is questionable whether the bill will actually be implemented.

Cyber-frauds have yet to reach their full potential on mobile platforms. Currently, the two most prevalent forms of mobile fraud are "cloning" and "scams." Cloning, which has become less common given security increases, involves a fraudster illicitly appropriating another user's mobile phone airtime. Prior to the digitization of mobile phones, this type of fraud was easily accomplished. Currently, mobile phone users are more likely to be subject to scams in which they are persuaded via a myriad of ways to dial a premium-rate number. More recently "smishing," a variant of phishing for mobile phones, has emerged: unsuspecting users are sent an SMS (short message service) message inviting them to click on a link. Once activated, all confidential details on the phone are downloaded for use by the fraudster. Such scams are currently in their infancy, but it seems inevitable that both domestic and business users alike will be increasingly targeted in attempts to appropriate proprietary information for criminal gain.

Cyber-Violence

Cyber-violence is the term used to describe online activities that have the potential to harm others via text and other "digital performances." These activities manifest in visual and audio forms, meaning that the violence is not actually physically experienced. Cyber-violence can be delineated by its perceived seriousness. Least serious are heated debates on message boards, often referred to as flaming (Joinson, 2003). At worst, a defamatory remark may be made about someone's inferior intellect or flawed argument. These exchanges are considered minor due to the fact that their consequences never amount to anything more than a bruised ego.

More serious are digital performances that are hate motivated. To take two examples, racial and homophobic hate-related online violence are in abundance in the form of extremist Web sites (Mann, Sutton, & Tuffin, 2003). Protected under freedom-of-speech laws in the United States, these sites employ shocking tactics to drum up support for their extremist viewpoints. In particular, some sites go as far as to display images of hate-related homicide victims in distasteful ways to heighten support for their very often misguided outlook on society's minorities (Schafer, 2002). The use of derogatory homophobic and racist text in these sites, combined with the use of inappropriate imagery and sound, results in a digital performance that is violent and potentially psychologically harmful, not only to the victim's family but also to the wider community.

Of potentially more harm are the violent activities of online stalkers. Cyber-stalking involves the use of electronic media, such as the Internet, to pursue, harass, or contact another in an unsolicited fashion (Petherick, 2000). Most often, given the vast distances that the Internet spans, this behavior may never manifest itself in the physical sense, but this does not mean that the pursuit is any less distressing. Wayne Petherick (2000) states that "there are a wide variety of means by which individuals may seek out and harass individuals even though they may not share the same geographic borders, and this may present a range of physical, emotional, and psychological consequences to the victim" (p. 1). Yet there still remains some concern that cyber-stalking might be a prelude to its physical manifestation (Reno, 1997).

The Internet allows communication with another person unconstrained by social reality, thus creating a certain psychodynamic appeal for the perpetrator who chooses to become a cyber-stalker (Meloy, 1998). Only written words are used, and other avenues of sensory perception are eliminated; one cannot see, hear, touch, smell, or emotionally sense the other person. There is also, if one wants, a suspension of real time. Messages can be sent and electronically stored, and their reception is no longer primarily dictated by the transport time of the medium, but instead by the behavior of the receiver. J. Reid Meloy (1998) explains that "some individuals may always return their phone calls the day they receive them, while reviewing their e-mail at leisure" (p. 11).

Meloy contends that these unusual circumstances provide opportunities for the stalker and presents a series of suppositions concerned with the medium itself. First, Meloy notes how the lack of social constraints inherent in online communication means that potential stalkers become disinhibited. Therefore, certain emotions and desires endemic to stalkers can be directly expressed toward the target more readily online than offline. Second, while online, the absence of sensory-perceptual stimuli from a potential victim means that fantasy can play an even more expansive role as the genesis of behavior in the stalker.

A more contentious debate in cyber-violence literature exists around the phenomenon that has become known as "virtual rape" (MacKinnon, 1997). These cases of virtual violence have completely escaped any legal rationalization. In the most famous case, a hacker was able to enter an online community and take control over community members' actions (Dibbell, 1993). Because movement and action within virtual communities are expressed through text, the "virtual rapist" was able to manipulate people's actions against their will. What essentially followed was a salacious depiction of violent rape upon several individuals in real time. While no one was physically harmed, members of the online community reported being traumatized by the event. This case was taken so seriously that the whole community (over a thousand members) voted on what action should be taken against the perpetrator. The imperative point to be made here is that the physical self of the perpetrator—the individual who exists in the offline world—could not be harmed; only his online persona could be punished. The effectiveness of such punishment is then questionable.

Mobile phone technology is not nearly sophisticated enough yet to facilitate the esoteric type of cyber-violence just discussed. More likely are acts of mobile stalking and bullying. The use of the Internet by pedophiles to groom children is well documented (see Quayle & Taylor, 2005; Taylor & Quayle, 2003). Mobile phones are the most recent addition to the forms of technology these individuals are employing to groom and stalk their victims. Often, initial contact is made via chat rooms or newsgroups. Communication then migrates to mobile platforms, allowing the perpetrator to maintain almost constant contact with the victim. Grooming via SMS text messages and phone conversations brings the pedophile psychologically closer to the victim, increasing the likelihood of a physical meeting. The near ubiquity of mobile phone ownership among young people in many industrialized countries increases the chances of mobile grooming and stalking to unprecedented levels (Sorensen, 2006). Young people themselves are also engaging in forms of cyber-violence in the shape of mobile bullying. In the recent NCH (2005) survey, 14 percent of respondents reported having suffered bullying via SMS messages on their mobile phone. This figure was three times that of those reporting bullying online. In the most extreme cases, victims who have suffered a torrent of mobile bullying attacks have been driven to suicide (NCH, 2005).

Regulating Static and Mobile Cyber-Crimes

Many attempts have been made to regulate the Internet. Acts of national and supranational governments on both sides of the Atlantic have attempted to criminalize various Internet-related activities, such as the posting of obscene and hate-related materials, the copying and distribution of intellectual property, and the grooming and stalking of children and adults. Specifically, failed attempts in the United States to regulate obscene online materials received international attention (such as the Berman Bill of 2002 and the Communications Decency Act of 1996). European legislation, including the EU Action Plan on Promoting the Safe Use of the Internet of 1997 and the Convention on Cybercrime of 2001, has been met with less contention. The latter is the most comprehensive piece of legislation that aims to criminalize and regulate certain behaviors online. Its reach is international, and it has been signed by the majority of the member states in the European Union and by the United States. While the act does not reference mobile platforms specifically, it defines cyber-crimes in a flexible way, allowing legal interpretation to include a range of existing and emerging technologies in their commission. The convention currently covers four areas of cyber-criminal activity: offenses against the confidentiality and integrity of computer data and systems; computer-related offenses; content-related offenses; and offenses related to infringements of copyright and related rights. Further additions to the convention concerning xenophobic content have also been included. While it is likely that the convention will stand the test of time and technology change, there are some doubts as to its successful implementation. For example, a major hurdle to the full incorporation of the convention into U.S. federal law concerns the protections afforded under the First Amendment.

Conclusion

The ability of cyber-crime to transcend place and space is further exacerbated by the migration to mobile platforms. There is little doubt that as communication technologies become progressively more mobile, cyber-crime will follow suit. The traditional threats to PC users—hacking, cyber-fraud, malware attacks, and cyber-violence—have been reengineered to function in the increasingly mobile world. However, the novelty of these mobile risks precludes mobile phone and wireless network users from accurately assessing the threat each time they connect. While corporate entities develop fixes to some of the threats (such as digitizing networks to reduce cloning), technology alone is not the answer to crime reduction. Crimes, including those that manifest on communication networks, are a social product. In tandem with technology, regulators must incorporate the social and market forces in the fight against what is doubtless to become the next generation of cyber-crime.

References

Association for Payment Clearing Services (APACS). (2004). Retrieved from http://www.apacs.org.uk/.

Aust, R., & Simmons, J. (2002). *Rural crime in England and Wales*. London: Home Office.

Becker, H. (1963). *Outsiders: Studies in the sociology of deviance*. New York: Free Press.

British Broadcasting Corporation. (1999). Soccer hooligans organise on the net. Retrieved from http://news.bbc.co.uk/1/hi/sci/tech/414948.stm.

Chalmers, R. C., & Almeroth, K. C. (2004). A security architecture for mobility-related services: Special issue on security for next generation communications. *Wireless Personal Communications*, 29(3/4), 247–261.

Computer Economics. (2005). Malware report: The impact of malicious code attacks. Retrieved from http://www.computereconomics.com/article.cfm?id=1090.

Department of Trade and Industry. (2006). Information security breaches survey. Retrieved October 6, 2006, from http://www.pwc.com/uk/eng/ins-sol/publ/pwc_dti-fullsurveyresults06.pdf.

Dibbell, J. (1993). A rape in cyberspace; or, How an evil clown, a Haitian trickster spirit, two wizards, and a cast of dozens turned a database into a society. *Village Voice*. Retrieved August 8, 2000, from http://www.levity.com/julian/bungle.html.

Gardner, W. (2003). The sexual offenses bill: Progress and the future. Paper presented at the Tackling Sexual Grooming Conference, Westminster, London. Retrieved from http://www.childnet-int.org/downloads/online-grooming2.pdf.

Gartner Research. (2004). Phishing attack victims likely targets for identity theft. Retrieved from http://www.gartner.com/DisplayDocument?doc_cd=120804.

Giddens, A. (1990). *The consequences of modernity*. Oxford: Polity Press.

Joinson, A. N. (2003). *Understanding the psychology of the Internet*. New York: Palgrave Macmillan.

MacKinnon, R. C. (1997). Virtual rape. *Journal of Computer Mediated Communication*, 2(4). Retrieved August 16, 2000, from http://jcmc.indiana.edu/v012/issue4/mackinnon.html.

Mann, D., Sutton, M., & Tuffin, R. (2003). The evolution of hate: Social dynamics in white racist newsgroups. *Internet Journal of Criminology*. Retrieved April 12, 2003, from http://www.internetjournalofcriminology.com/ijcarticles.html.

McCormack, M. (1996). Computer hackers turn to pager systems. *Computers and Security*, 15(7), 585–586.

Meloy, J. R. (1998). The psychology of stalking. In J. R. Meloy (Ed.), *The psychology of stalking: Clinical and forensic perspectives* (pp. 2–27). London: Academic Press.

NCH. (2005). *Putting u in the picture: Mobile bullying survey 2005*. London: NCH.

O'Malley, K. (1995). Costly cellular phone fraud rises with number "cloning." *Computers and Security*, 14(2), 119.

Petherick, W. (2000). Cyber-stalking: Obsessional pursuit and the digital criminal. Retrieved May 2, 2001, from http://www.crimelibrary.com/criminal_mind/psychology/cyberstalking/1.html.

Ponemon Institute. (2004). *National spoofing and phishing study*. Retrieved from http://www.ponemon.org/emerging.html.

Quayle, E., & Taylor, M. (2005). *Viewing child pornography on the Internet: Understanding the offense, managing the offender, helping the victims*. Dorset: Russell House.

Reno, R. Hon. J. (1997). Keynote address to the meeting of the G8 Senior Experts Group on Transnational Organized Crime, Chantilly, VA. Retrieved March 20, 2002, from http://www.usdoj.gov/criminal/cybercrime/agfranc.htm.

Richardson, R. (2003). *CSI/FBI computer crime and security survey*. San Francisco, CA: Computer Security Institute.

Schafer, J. A. (2002). Spinning the web of hate: Web-based hate propagation by extremist organizations. *Journal of Criminal Justice and Popular Culture*, 9(2), 69–88.

Sorensen, C. (2006). *The mobile life 2006: How mobile phones change the way we live*. London: Carphone Warehouse.

Sussmann, M. A. (1999). The critical challenges from international high-tech and computer-related crime at the millennium. *Duke Journal of Comparative and International Law*, 9, 451–489.

Taylor, M., & Quayle, E. (2003). *Child pornography: An Internet crime*. London: Routledge.

Telecommunications, Network and Security Research Group (TNS). (2006). *Mobile Internet survey*. London: TNS.

Wall, D. S. (1998). Policing and the regulation of cyberspace. [Special edition on crime, criminal justice, and the Internet]. *Criminal Law Review*, 79–91.

———. (2001). Cybercrimes and the Internet. In D. S. Wall (Ed.), *Crime and the Internet* (pp. 1–17). London: Routledge.

Williams, M. (2006). *Virtually criminal: Crime, deviance and regulation online*. London: Routledge.

· 7 ·

BREAKING FREE

The Shaping and Resisting of Mobility in Personal Information and Communication Technologies

Julian Kilker

The iPod is a very nice device with a great UI [user interface]—but if there's something about it that you don't like, then why shouldn't you be empowered to change it?

PCM2, posting to technology discussion site Slashdot.org, April 14, 2006

Ever since writing was invented to keep records and communicate messages over great distances, information and communication technologies (ICTs) have modified how people have interacted over time and through space. In the early twenty-first century, portable on-demand (POD) devices, such as cellular phones with data capabilities, Internet video, voice over Internet protocol (VoIP), digital media players, and receivers that digitally record programs for later consumption, have allowed us to manipulate time and space far beyond earlier technologies. Such POD innovations are possible because of technical advances in microprocessor, memory, and battery technologies that have reduced the cost and weight of transporting information, and enabled technologies such as global positioning system (GPS) devices and mobile phones to be "location aware." The fact that many ICTs were designed to be flexible so that their developers, and to some extent users, could modify POD characteristics has received little attention in the popular press and research literature. Such modifications can change which activities (communication, viewing, recording, editing) can take place where (at home, work, while traveling, wirelessly); essentially, they can change how "place" is defined in communication terms. Moreover, as this chapter will highlight, producers and users have often had different expectations and goals regarding the manipulation of place. User interfaces are the reification of designers' decisions about user control; such decisions concern not only specific control characteristics but also whether there are options for users' reconfiguration. This distinction between basic control and the reconfigurability of control is absent in most discussions of the interactivity of ICTs.

The specific POD technology examined in this chapter is the digital media player—specifically, the operating system, or firmware, that controls such players. Digital media players have rapidly become popular since their introduction to the U.S. mass market in the winter of 1998 by manufacturers such as Diamond Multimedia and Creative Labs. The Apple iPod, with its companion iTunes software that enabled the purchasing, downloading, and organization of audio content (and video content in later versions), was a late starter in 2001, but has become the dominant brand because of both the seamlessness of the technology's user interface and the sophistication of Apple's marketing. Each release of a new iPod generation receives media publicity far more extensive than that of other manufacturers, effectively making the iPod the standard by which other players are judged. (When other players receive media attention, the words "iPod killer?" are often used, although the answer is usually negative.) In October 2006, Apple introduced iPod generation "5.5," which upgraded the video-playing capabilities and enabled legal download of movies, among other features. The word *iPod* has even become a generic name for digital media players, despite the wide variety of players on the market that differ more in terms of industrial design and user interface than in functionality.

Portable media players are influential both commercially and socially. Commercially, they appear to have played a large role in the decline of radio listeners, especially among the young. According to Arbitron, the time spent listening to radio from spring 1999 to spring 2006 was down 15 percent, 15.3 percent, and 13.2 percent in the twelve-to-seventeen, eighteen-to-twenty-four, and twenty-five-to-thirty-four age groups, respectively; overall, those aged twelve to thirty-four listened to seventeen hours of radio a week in spring 2006, three hours fewer than in spring 1999 (Siklos, 2006). From a social perspective, research has shown that listeners use portable players to manage space and time, and to control interactions with others (Bull, 2000). But rather than focusing on how portable on-demand devices such as the iPod are shaping society—a perspective well represented in both the popular and research literature—this chapter examines how people (re)construct media technologies to alter their own media environments. Nelly Oudshoorn and Trevor Pinch (2003) have argued that "users and non-users matter," as well as the producers of technologies. This chapter argues that people who are technology "modders" (modifiers)—positioned between users and producers—are also of great importance.

Fundamentally, the user-generated modifications for many consumer ICTs demonstrate that both users and producers are "breaking free": user-modifiers are attempting to break free from limitations embedded in the technologies, while many producers are attempting to break what they perceive as threats to their business models in which the modifications add value for free. The actions of both groups have implications for how "mobility"—and its associations for modifying notions of place and time—is reified in portable on-demand technologies.

In this chapter, the essential foundations of collaborative technology development are discussed first. The collaborative modification of ICT-based media systems began with the ARPANET computer research network; this sociotechnical context has strongly influenced most recent media systems. Then the "meta-control" perspective is introduced to explain notions of control in media technologies. (I have used this perspective in an analysis of streaming media and DVD players [Kilker, 2003] and in a comparison of classroom- and Web-based pedagogy [Kilker, under review]). Finally, I present a case study of Rockbox, a user-developed software that modifies the operation of several different portable media players, including the iPod. The Rockbox project and its consequences demonstrate the collaboration, exploration, and control issues that are at the heart of how "breaking free" affects place.

This chapter, consistent with the theme of this entire volume, is an examination of the relation between ICTs and space prompted by Joshua Meyrowitz's seminal *No Sense of Place* (1985), written before the Internet was available to the broad public in the early 1990s. In the book, Meyrowitz attempts to explain changes in society at the time through a synthesis of Marshall McLuhan's (1964) insights on the effects of electronic media on society and Erving Goffman's (1959) analysis of social roles and identities. Meyrowitz's argument is largely technologically deterministic; his book's subtitle is "The Impact of Electronic Media on Social Behavior." In discussions of technology and society, the influence of typical users on technology had often been neglected; in response, perspectives such as "co-construction" and "mutual shaping" have been developed to examine interactions between the two. This chapter uses the meta-control perspective, which emphasizes the mutual shaping perspective and focuses on the ability to reconfigure interactivity control options available to a technology's users (Kilker, 2003).

Meta-control is a useful concept in media contexts because it describes how people are exposed to, interact with, and influence media systems, particularly media systems that rely on ICTs, such as online video and DVD players. Because meta-control shapes people's ability to interact with media, it influences the characteristics and perceptions of the medium. Examining control characteristics can lead to a better understanding of a medium's adoption and resistance by multiple social groups; in the case of POD devices, these social groups include developers, content providers, technology "hackers," and general users. Fundamentally, the meta-control perspective emphasizes the social expectations embedded in technology.

POD technologies are hybrid systems of software, hardware, and personal and network technologies, and their capabilities can be controlled through programming and user preferences. The extent to which user control of new media technologies can be modified depends on whether the technology supports reconfiguration, typically through software, which—unlike hardware—is inherently reconfigurable. Some POD technologies, such as iTunes (when operated on a notebook computer), are entirely software functioning on generic personal computer hardware, and thus can be easily modified by software updates or patches.

The implications of the convergence of ICTs were identified in detail in an early technical article by computer scientists J. C. R. Licklider and Robert Taylor (1968), who were influential in the development of ARPANET, the model for the present-day Internet. These researchers reported using time-sharing computers to collaborate with others; the illustrations in the article show a group of managers using typewriter-sized computer terminals as "personal digital assistants" (PDAs) in a conference room. (These devices provided information on demand, but, unlike today's PDAs, they were certainly not portable.)

A prescient conclusion of this 1968 article on social interactions, which appeared in the technical journal *International Science and Technology*, was that special-interest communities would likely form using similar technologies. Shortly thereafter, the first ARPANET connections were constructed at several locations across the United States. Significantly, the ARPANET was not only an experiment in technical "interoperation" (meaning that the computer technologies had to work together smoothly), but also an attempt to encourage social interoperation (collaboration) and to build computer science as a field through the funding of key resources in separate institutions. As discussed in my research on the development of technical standards for networked email, the ARPANET's design was not only a remarkable technical feat in which participants continually reconstructed and optimized their communication tools; it was also a remarkable feat of social collaboration in which the design values, goals, and scenarios of different social groups were negotiated (Kilker, 2002).

Why is this historical contextualization important? The "electronic media" that Meyrowitz (1985) and McLuhan (1964) analyze are television and radio, not computer networks. Both television and radio were hardwired, centralized, and homogenous media systems in which the audience had little influence over the underlying medium. Meyrowitz did note that "uses and gratifications" research assumes a more active role for audience members. Subsequent research has examined the idea that people can partially control video media, using, for example, the remote control to "zap" (change channels) and "zip" (fast-forward) to avoid advertising (Bellamy & Walker, 1996). However, for the most part people had no control over media technologies themselves. In the pre-ICT era, modifying media technologies required opening radio and television cabinets, disregarding the "Warning: High Voltage!" labels; even hobbyists assembling their Heathkit televisions were replicating, not modifying, existing designs. (From the 1940s to 1980s, Heathkit provided kits for electronics hobbyists that were famous for their solid design and detailed assembly manuals.)

The ARPANET era introduced three key innovations that influence the present-day design of ICTs: The first innovation is that an ICT-based media technology is malleable. The second is the notion that improvised development and collaboration—rather than formal institutional structures—can produce technologies that reflect the skills and needs of a broader range of people. Douglas Engelbart's research group at the Stanford Research Institute developed the design philosophy of "bootstrapping" (from "pulling one's self up by the bootstraps"), which espouses that designers' decisions should be actively influenced by their own use of the technology (Bardini & Horvath, 1995). The essential third innovation is that, because of its flexible design, changes to the ARPANET were easily propagated across the network, even to locations where people had limited computer expertise. People did not have to open hardware and expose themselves to dangerous voltages to reconfigure the network; instead, they installed software that contained in itself both the modification and the "expertise" needed to install it; indeed, installing software is now familiar and deceptively unremarkable to the many people who regularly download and install software that modifies the capabilities of their personal computers.

The Meta-Control Perspective

User control has been popularized by works such as *The Design of Everyday Things* (Norman, 1988), which analyzes how people interact with commonplace technologies such as doors and cars. Consumer devices that rely on ICTs have become similarly "everyday," and their design and use share similar control factors. ICTs allow interactivity characteristics to be shaped, selectively restricted, and resisted, amplifying the tensions among the groups of people associated with the technology. Each POD media device, which merges communication and information technologies, is the result of a negotiation of different design perspectives based on the hybrid ancestry of mass media and computer science, which have fundamentally different concerns about control.

The distinction between basic control of an ICT and the ability to reconfigure control is important. Whereas a person usually controls the basic features of a technology (those defined in the user's manual), and may even be able to reconfigure these features in a limited fashion using customization and personalization tools, the ability to define the boundaries of these features is usually limited to the technology's producers, as well as to advanced and determined users. The Internet has facilitated the transmission of information, both social and technical, that enables the reshaping of POD devices, just as the ARPANET facilitated the development and shaping of its key services, such as email.

Meta-Control and Place

While technologies can control media consumption through the use of meta-control limitations, research on the social shaping, domestication, and reinvention of technologies in general suggests that users will attempt to use media in ways unintended by their original producers.

Consider the following ICT media technologies that manipulate time and space, sometimes in ways that are indirect. The Razr mobile phone, a collaboration between Motorola and Apple, features a digital music player. To protect Apple's and the phone providers' revenue streams, the number of audio tracks that the Razr can hold is limited to one hundred, and the audio tracks cannot be used as ring tones. Similarly, in order to license DVD technology, producers of DVD players are required to implement restrictions that limit viewer control over video discs, including when viewers can fast-forward past content (called "UOP," for "user operation prohibition") and in which geographic regions discs can be played (called "RPC," for "regional playback control"). DVD players are coded by the region in which they are ostensibly sold (region 1 for the United States, Bermuda, Canada, and U.S. Territories; region 2 includes Japan, Europe, Greenland, South Africa, the Middle East, and Egypt, and so on), and video discs are coded to allow (or not allow) playing in each region. This is designed to enable producers to control the distribution of content. In practice, many reasonable exceptions exist for using "out of region" content, and region coding technology is readily resisted and disabled. Mixed experiences with DVD region coding has led to intense interest in how region coding will be embedded and used in HD-DVD content.

As this chapter was being completed, new cases of media control continued to emerge. In October 2006, Creative released a firmware upgrade for its Zen MicroPhoto and Zen Vision:M media players that added some features, fixed a few bugs, but also, surprisingly, removed a favorite feature that allowed digital recording of FM radio content, apparently under pressure from the Recording Industry Association of America (Miller, 2006b). Amazon launched Unbox, a video-on-demand service designed to compete with Blockbuster and Netflix, in September 2006. Ultimately, according to early strongly negative reviews (Doctorow, 2006; Lewis, 2006), Unbox content seems less convenient and portable than DVDs for reasons of control (the ability to navigate content is limited), licensing (content cannot be legally viewed in many locations, including outside the United States and in "hotel rooms, motel rooms, hospital patient rooms, restaurants, bars, prisons, barracks, [and] drilling rigs"), privacy (the software apparently reports information to Amazon about viewers' computer and viewing habits), and pricing (in some cases, the on-demand content cost more than a physical DVD).

Also in September 2006, Microsoft announced details about its Zune portable media player, designed to compete with Apple's iPod-iTunes music juggernaut. A major feature is its ability to share media files with nearby Zune users, but under restrictive circumstances designed to encourage their purchase. As a highly critical CNET article notes, "But all things must pass—in this case, within 72 hours. You'll have three days in which to listen to each song three times. . . . Doesn't matter if you're passing your parents a recording you made of your kid being cute—Zune will banish it" (CNET News, 2006a).

Because the Zune cannot determine whether the file being shared is a commercial music track, your garage band, a lecture recording, or a child's first words, it subjects all shared audio to the most extreme digital rights management control. These arbitrary limitations embedded in Zune's design not only contradict how the "smart kids" use media today, but also challenge them to resist such restrictions: "you can see them listening to radio over their wireless networks (Zune says no), picking up streaming media from blogs (Zune says no), pumping in DivX TV programmes grabbed off the torrents (Zune says no, no, no)" (CNET News, 2006a).

Control limitations such as these in the Razr, DVD players, and the Zune are not the focus of mainstream media reviews, but they are emphasized by many online reviews and discussions. They are certainly not emphasized in advertisements, which typically laud interactivity and other technological features. Yet these limitations have consequences for the success of the projects, and for how people resist them.

Many tactics have emerged to address limits on what people can watch, and where. DVD players have been modified and software is widely available to disable "region coding" and reestablish the viewer's control over place. The regional blanking of broadcasts has been addressed by the Slingbox, a device that transmits video over the Internet from a person's home cable box, so that a traveler can use a laptop computer to view television programming as if at home. Similarly, when people were dissatisfied with their local broadcast offerings of FIFA's World Cup 2006—game matches were selectively broadcast and time-shifted—some turned to online resources and used the peer-to-peer viewer TVUPlayer to view matches live from other continents.

Modding Communities

Flexibility in information technology standards, openness of collaboration, and the ability to modify technologies via easily propagated software, merged with an interest in understanding and modifying technology, sets the stage for the present-day "modding communities," groups of people who collaborate online in developing and propagating modifications to existing ICTs. It might seem peculiar to focus on such communities of technology modifiers. After all, craftspeople and hobbyists have long modified technological artifacts on their own, and their influence beyond their immediate surroundings seems minor, at best.

However, two arguments challenge this. First, online modding groups have clearly been influential. Early adopters of technologies tend to be particularly influential in their recommendations and concerned about their ability to control and explore these technologies. In the case of media technologies, as described earlier, people are often frustrated with how ICTs limit place-shifting and time-shifting. Frustration with similar limitations has led people to discuss and collaborate online in the development of control solutions. For example, fans of the Complete New Yorker archive, which stores all issues of the magazine on several discs that must be frequently swapped in and out of a computer, publicized techniques to load all of its issues on a hard drive for portability and speed. DeCSS, software written by only three people (two anonymously) to decrypt DVDs so that they could be copied, is at the core of all software that decodes and transfers DVD content to other players, including the Sony PSP and laptops. (Although it can be used for illegal purposes, DeCSS also has ethical applications, such as playing DVDs on the Linux operating system and creating personal backups of DVDs. However, in the United States, the DeCSS algorithm is illegal to publish as a result of the Digital Millennium Copyright Act, a coup for content producers in meta-control terms [Lessig, 1999]). In addition, the importance of studying the "alpha geek" has been demonstrated by technology publisher O'Reilly's popular Emerging Technology conferences, at which advanced early adopters of technologies discuss trends with each other and less-advanced participants.

Second, earlier versions of modding communities have been studied by researchers. The research frameworks have included: "domestication" (an approach emphasizing the integration of technological artifacts into everyday, usually household environments [Silverstone & Haddon, 1996, p. 44]); "reinvention" (a modification of the diffusion of innovations model that acknowledges influences on the innovations by their adopters [Rice & Rogers, 1980]); and "co-construction" (examining user-technology relations [Oudshoorn & Pinch, 2003]).

In contrast to the typical popular accounts of technological development, which emphasize institutional factors and individual inventors, historians of technology have demonstrated how early users were extremely influential in determining not only the uses for technologies, but also their technological configurations. For example, amateurs developed unanticipated uses for early radio that were then adopted by commercial users (Douglas, 1987); rural U.S. farmers adapted their early gasoline vehicles to power a wide range of labor-saving devices, such as washing machines and plows (Kline & Pinch, 1996); and women used the early telephone to overcome social isolation—although it was initially framed by telephone companies as a commercial tool, like the telegraph—which led telephone companies to shift their business models (Fischer, 1992).

In the early amateur radio and farming examples, the people modifying the technologies collaborated through advertisements and articles in publications available to their communities. Social collaboration resources available on the Internet now facilitate the formation and coordination of special-interest groups. While technically adept users typically initiate the shaping of ICTs, their instructions and modifications can be transmitted over computer networks for implementation by relative amateurs. Thus, modding is greatly facilitated by the Internet, which has a long history of enabling early adopters of technologies to organize, form special-interest groups, and disseminate both tips and technological modifications (if they are software based) that can shape ICTs.

A wide variety of modding sites exist online, with important implications. Their content is readily accessible via popular search engines such as Google, which allows nonexperts to stumble across these groups when searching for information about a product. In addition, easy access to the groups allows prominent and high-traffic Web sites, such as Slashdot (www.slashdot.org) and Boing Boing (www.boingboing.net), to reference them periodically. By publishing to a large number of people, such sites amplify the effects of personal projects in which advanced users have shaped their ICTs. The amplification effect is furthered by technology writers who regularly scan sites such as Slashdot for technology trends and then report on them.

Key goals of modding communities are to explore, understand, and control technologies. As PCM2, the person quoted in the introduction, notes, "This idea that products arrive at your door like the manufacturer built them and you do what the manufacturer wanted you to do with them and when you don't like it anymore you throw it away is fundamentally f'ed up. It's a product of the modern disposable consumer culture that, if you're really a geek, you should be fighting against" (PCM2, April 14, 2006; see Newman, chapter 5 in this volume).

Modding is especially popular with video game players, in part because it is often supported by the game developers, who release development tools. Some "mods" are relatively simple, in that they modify the "skin" (the visible, surface elements) of the technology rather than the underlying technologies themselves. (Software programs often allow users to modify the "skin" or look of an application's user interface, but not its underlying control characteristics. Similarly, there are thriving online communities that focus on "case mods" for computer hardware.) But volunteers have also created complex modifications of many popular games such as Half-Life 2, including entirely new game levels (such as Adam Foster's highly regarded Minerva) and "sandbox" experimental environments in which players can position and test game elements (such as Garry Newman's Garry's Mod). While the utility of this sandbox resource might not be obvious, Garry's Mod has enabled other modders to create a wide variety of mini-games within the Half-Life 2 environment, and it has enabled the author of the webcomic parody *Concerned* to pose characters from the game to create "snapshots" for his illustrations.

Most major computer games have mods and groups that support them; see Wikipedia's "mod (computer gaming)" entry for a typology of mods and an extensive list of examples. In addition to the popularity of computer game modding, there has been a recent resurgence in the popularity of modifying consumer technologies, facilitated by the exchange of information online and resources such as O'Reilly's *Make* magazine and Web site, which were started in 2005, and books such as the insightful *Hacking the Xbox* (Huang, 2003), which discuss hacking technologies in the original exploratory (rather than destructive) sense of the term.

Rockbox Modding Community Case Study

The Rockbox modding community demonstrates meta-control in and shaping of POD media technology. Rockbox is an open-source operating system for portable media players designed in response to user frustrations with the limited and unreliable software that shipped with the original Archos Recorder in 2001. Rockbox now works on several different players, including iRiver models (in 2004) and various iPod models (in 2005), although at different levels of compatibility.

POD devices are controlled by software that mediates between the user and the technology. Rockbox takes over control of the player, replacing the manufacturer's operating system with an expandable system. An introductory page on the project Web site (www.rockbox.org) entitled "Why Rockbox" notes that "Rockbox aims to be considerably more functional and efficient than your device's stock firmware while remaining easy to use and customizable. Rockbox is written by users, for users." It continues, "A major goal of Rockbox is to be simple and easy to use, yet remain very customizable and configurable. We believe that you should never need to go through a series of menus for an action you perform frequently. We also believe that you should be able to configure almost anything about Rockbox you could want, pertaining to functionality."

In the case of the original Archos device, the new Rockbox operating system dramatically improved the quality of the user interface; in other digital media players with more powerful technology, Rockbox was able to provide additional capabilities but at the expense of complete compatibility. For example, on the iRiver model, Rockbox added features such as gapless playing (before the iPod did in 2006), FM recording, custom user interface themes ("skins"), additional digital file formats, voice user interface (for vision-impaired users), and simple games and accessories. It also fixed bugs in recording and playlist shuffling. On the iPod version, Rockbox adds similar features (although because it also explicitly contains no digital rights management, iTunes-purchased files cannot be played).

Rockbox has been collaboratively developed by a core group of people using the Concurrent Versioning System (CVS), which tracks versions of software files and is used to coordinate many software projects. The most recent CVS modifications are listed on the main page of the project's site, with the changes and the programmer's name attached. The core developers are based in Sweden, but the main language of the site is English. According to self-reports on the site, as of September 18, 2006, there were seventy-five developers across Europe, North America, and Australia, with single developers in Japan and Colombia. The site also reported approximately fourteen hundred users throughout the world, with large concentrations in Europe and the United States but many others in developing countries as well. (Because this self-reporting feature had been active for less than one month at the time I recorded these figures, they are probably underestimated.) In a 2006 online interview, Daniel Stenberg, one of the three original developers, estimated that there were approximately "180 named contributors, ~3000 registered users in the Rockbox forum, ~1000 subscribers to the mailing lists and a few hundred registered users in the bug tracker, but regularly . . . hard to tell. We're some 30 people that are more or less 'core' people" (Marinho, 2006).

Most of these people have not met in person, although there was a Developers Convention in Stockholm in March 2006 that attracted eight core developers. The online invitation noted that "Since we can code and IRC at home, too, we should focus on helpful group activities" and listed items to negotiate in person on the attached agenda.

Other areas of the Web site list current feature requests and bug reports from users. Background information on each digital media player summarizes what developers have learned: which processor is used, how the hardware works, what techniques to use to load test software and recover from the inevitable crashes, circuit board pictures showing important components, and current open questions about the player technology. Another area of the site archives (from 2002 to the present) the Internet Relay Chat (IRC) channel devoted to Rockbox discussions. For example, on September 18, 2006, "<desrt>" asks, "does anyone know what sort of cpu is in the new ipod nano?" and "<preglow>" answers, "a samsung arm/calmrisc affair." In other words, a Samsung RISC processor drives the Nano. This is information developers need to understand the capabilities of the technology and whether Rockbox can operate on it.

The technical specifications and discussions on this site form a type of collaborative volunteer software development laboratory "by users, for users," controlling their POD media experiences. The core developer, Stenberg, noted that "it's always very rewarding to consider that we are now actually outperforming lots of original firmwares," but that the modding group's official interaction with the manufacturers of the various players had been limited. The two exceptions are that "Archos . . . wanted to ship Rockbox on some sort of CD with their players, but since they wanted to say it was a 'cooperation' between them and us we rejected that suggestion since they never helped us one tiny bit" (Marinho, 2006), and that SanDisk, as part of its anti-iPod marketing campaign, had apparently contacted Rockbox developers to request support for SanDisk players (CNET News, 2006b).

Another section of the site lists Rockbox's media coverage. All of the references are from online sources. *PCMag*'s article is entitled "Install a Second Operating System to Unveil Hidden Power." A sample quote is: "Are you dissatisfied with your iPod's standard capabilities? Why not install one of the two major alternative operating systems: iPodLinux or Rockbox" (Kobrin, 2006). *Engadget*'s June 2, 2006, article is entitled "SanDisk Hits up Rockbox for Some Firmware." A sample quote is: "They've just pinged the Rockbox community for a port of the open source Rockbox firmware . . . , which at least should give them some serious nerd cred in a time when most everyone else is locking down their hardware" (Miller, 2006a). Reference to reputable external sources such as these on the Rockbox site reassures potential users that the project is legitimate—and thus that the free software does not contain "malware" (see Williams, chapter 6 in this volume)—and assists them in evaluating whether to adopt it themselves.

The Rockbox project was begun because of frustrations with one player, but now it can be installed on multiple models from four manufacturers (Archos, iRiver, Apple, and iAudio), with other versions in development. As is apparent from the overtures from Archos and SanDisk, as well as recent feature upgrades to Apple's iPod, the Rockbox project has exerted competitive pressure on commercial manufacturers.

From this brief overview of the Rockbox project, it is clear that people can and do collaborate to modify portable digital technologies, in effect exerting meta-control over how these technologies operate, which activities people can do with them, and where they can engage in these activities. In summary, the portable digital media player, typical of many recent convergence technologies, is usually designed to allow software updating. The shortcomings of commercial versions, coupled with the potential of the hardware, have attracted a group of people interested in modifying the technology to fix problems, add features, and increase flexibility. These people, who constitute a "modding community," are dispersed across multiple continents and time zones, yet are able to collaborate effectively in researching and modifying the technology using Internet communication tools (IRC, a Web site hosting archives, and webcams) and specialized software designed for collaborative development (CVS). The modifications, packaged in the form of easy-to-install software, are readily distributed online. People installing the modifications can provide feedback, identify problems and desired features, and, if technically competent, implement changes online.

Conclusion

Studying emerging technologies such as POD media systems is important for understanding topics such as meta-control and modding communities. However, systematically researching these aspects of emerging media involves collecting and preserving as much information as possible. A key challenge in examining technological artifacts is their rapid change. Over the past several years, digital audio players have gone through several generations of development, from CD based (mechanical and quite large) to hard-drive based (palm sized and capacious, but fragile) to flash-memory based (light and rugged, but with limited space). While these players are featured in advertisements and reviews, modifications such as the Rockbox project have received little attention in the mass media, apart from online reviews. This creates a challenge for future researchers: consider that the thorough historical research on the modding of farmer's vehicles (Kline & Pinch, 1996) and the early radio (Douglas, 1987) depended on the availability in research libraries of farming and hobbyist publications that documented modifications. In fact, much historical research about technological development is biased toward commercial products whose companies are able to indirectly "purchase their place in history" by buying advertisements in publications that are eventually archived in research libraries. In contrast, the noncommercial Rockbox project is documented in non-mass media, and in particular online resources that might not be adequately archived or accessible in the future. This dilemma is present in this chapter: while the theoretical references are from traditional peer-reviewed academic sources, the primary source references for the emerging media examples and the Rockbox case study are from perishable online sources, some of which will likely be unavailable by the time this book is published.

This chapter presented the concepts of modding communities and meta-control, and explored their implications in shaping portable on-demand media technologies, such as the digital media player. While commercially available POD technologies are often framed in terms of enhanced flexibility in the manipulation of time and place, design restrictions have led to technologies that limit user control over such activities. Modding communities, such as the developers of Rockbox, attempt to expand users' ability to control such technologies, usually without support from their original producers.

The example of Rockbox demonstrates "displacing place" in two important ways. First, geographic location is not a key factor in the modification of the technology, as the core development group, and to a larger extent its users/testers, are geographically dispersed. Many examples of dispersed social groups exist, online and offline, but this type of informal placeless community has produced a functional artifact as a product of its collaboration. Second, the modifications to this POD media technology change what its users can do with it, and where they can do these things. For example, in the Rockbox case, people can now record and play games on the move, and have their device behave differently when operating in a car or while walking. Modifying a POD media technology effectively modifies how its users are tied to place.

An important implication of the convergence of information with communication technologies, in which many features are implemented in software rather than hardware, is the ability to modify control characteristics. The media examples presented here demonstrate that research on the emergence of new media and their characteristics should include a thorough analysis of how people influence the technologies in ways that range from direct modification to installed modifications acquired online from modding communities. Such research requires the documentation and preservation of information about such modifications, which tend to be overlooked by mainstream chroniclers, and an examination of technologies as products of multiple social groups, including producers, users, nonusers, and, now, modifiers as well.

References

Bardini, T., & Horvath, A. (1995). The social construction of the personal computer user. *Journal of Communication*, 45(3), 40–65.

Bellamy, R. V., & Walker, J. R. (1996). *Television and the remote control: Grazing on a vast wasteland*. New York: Guilford.

Bull, M. (2000). *Sounding out the city: Personal stereos and the management of everyday life*. Oxford: Berg.

CNET News. (2006a). Microsoft Zune: All the excitement that brown can bring. Retrieved September 15, 2006, from http://crave.cnet.co.uk/digitalmusic/0,39029432,49283610,00.htm.

———. (2006b). SanDisk goes after the iPod iPuppets. Retrieved September 15, 2006, from http://crave.cnet.co.uk/digitalmusic/0,39029432,49273890,00.htm.

Doctorow, C. (2006, September 15). Amazon Unbox to customers: Eat shit and die. Retrieved September 15, 2006, from http://www.boingboing.net/2006/09/15/amazon_unbox_to_cust.html.

Douglas, S. (1987). *Inventing American broadcasting, 1899–1922*. Baltimore, MD: Johns Hopkins University Press.

Fischer, C. (1992). *America calling: A social history of the telephone to 1940*. Berkeley: University of California Press.

Goffman, E. (1959). *The presentation of self in everyday life*. New York: Doubleday.

Huang, A. (2003). *Hacking the Xbox: An introduction to reverse engineering*. San Francisco, CA: No Starch Press.

Kilker, J. (2002). Social and technical interoperability, the construction of users, and "arrested closure": A case study of networked electronic mail development. *Iterations*, 1(1).

———. (2003). Shaping convergence media: "Meta-control" and the domestication of DVD and Web technologies. *Convergence*, 9(3), 20–39.

———. (Under review). Procrustean pedagogical environments: Understanding usability, interactivity, and control in Web-based course management systems.

Kline, R., & Pinch, T. (1996). Users as agents of technological change: The social construction of the automobile in the rural United States. *Technology and Culture*, 37(4), 763–795.

Kobrin, M. (2006, August 30). An Apple iPod: Install a second operating system to unveil hidden power. Retrieved September 20, 2006, from http://www.pcmag.com/article2/0,1759,2010245,00.asp.

Lessig, L. (1999). *Code and other laws of Cyberspace*. New York: Basic.

Lewis, P. (2006, September 19). Two thumbs down for Unbox: Amazon's new movie service is a horror show. Retrieved September 20, 2006, from http://money.cnn.com/2006/09/18/technology/lewis_unbox.fortune/index.htm.

Licklider, J. C. R., & Taylor, R. (1968, April). The computer as a communication device. *International Science and Technology*, 21–31.

Marinho, R. (2006, June 8). Rockbox: The open source jukebox firmware: The interview. Retrieved September 20, 2006, from http://www.cdfreaks.com/article/280/3.

McLuhan, M. (1964). *Understanding media: The extensions of man*. New York: McGraw-Hill.

Meyrowitz, J. (1985). *No sense of place: The impact of electronic media on social behavior*. New York: Oxford University Press.

Miller, P. (2006a, June 2). SanDisk hits up Rockbox for some firmware. Retrieved September 20, 2006, from http://www.engadget.com/2006/06/02/sandisk-hits-up-rockbox-for-some-firmware/.

———. (2006b, October 16). Creative removing FM recording from players. Retrieved October 18, 2006, from http://www.engadget.com/2006/10/16/creative-removing-fm-recording-from-players/.

Norman, D. A. (1988). *The design of everyday things*. New York: Basic Books.

Oudshoorn, N., & Pinch, T. (Eds.). (2003). *The co-construction of users and technology*. Cambridge, MA: MIT Press.

Rice, R., & Rogers, E. (1980). Reinvention in the innovation process. *Knowledge*, 1, 499–514.

Siklos, R. (2006, September 15). Changing its tune. *New York Times*, p. C1.

Silverstone, R., & Haddon, L. (1996). Design and domestication of information and communication technologies: Technical change and everyday life. In R. Mansell & R. Silverstone (Eds.), *Communication by design* (pp. 44–74). Oxford: Oxford University Press.

·8·

MOBILE CULTURE

Podcasting as Public Media

Jarice Hanson and Bryan Baldwin

An entire history of Federal Communications Commission (FCC) decisions and debates concerning the scarcity of airwaves and the need for alternative voices in the media has focused on whether the public is or is not exposed to a number of different viewpoints and values in different forms of public media. Since the Telecommunications Act of 1996, critics have pointed to changes in ownership caps that have effectively put more media content in the hands of a mere five conglomerates, resulting in fewer alternative voices, viewpoints, or values (see, for example, Bagdikian, 2004; and McChesney & Nichols, 2002).

Prior to the days of deregulation in the 1980s, the rallying cry of those who advocated "marketplace rules" was, in part, based on the claim that new media distribution forms and better use of the electromagnetic spectrum no longer required that the spectrum be "regulated" as a scarce commodity (Fowler & Brenner, 1988). When VHF and UHF frequencies were made available for television, and cable television emerged in the 1970s, subscription services began supplying special programming in the 1980s, and when the Internet started distributing enhanced audio and text files in the 1990s, advocates of "marketplace rules" shouted victory over earlier interpretations of the airwaves as a public trust, and freedom of expression and freedom of speech began to be regarded as the right of the media owner, rather than the right of the media audience.

In recent years, podcasting has also been described as an alternative distribution form for a multitude of small-format, personal information sources. Conceptually, podcasts are most similar to radio, but do not require the podcaster to be licensed, nor do podcasters use airwaves for the distribution of their signals. If they are most like radio, a reasonable approach to considering the format aspects of podcasting might be to acknowledge that radio has undergone a marked transition since the days in which it was the primary form of communication for a nation. It is generally agreed that radio in both the United States and Canada has increasingly shifted to one of two models: syndicated programming that is run by chain radio broadcasters, or local radio that continues to try to serve the needs of a geographical community. Because podcasting is such a personal medium and can only be accessed by personal technologies (most of which are portable), we chose to explore podcasting as a possible compliment or adjunct to local radio.

At first glance, podcasts and podcast technology seem to fit into the category of public access, and therefore could reasonably be expected to represent a form of public media. The portability of iPods and other technologies with downloading capabilities, such as cell phones and laptop computers, contributes to the small-format, mobile features that make podcasting possible, and makes podcasting a suitable topic for a study of mobile culture. After all, podcasting can free the listener from any fixed location, and podcasts can be heard when the listener has time to attend to the content.

Conceptualizing the Study

We began our investigation of whether podcasts did, indeed, represent a form of public media by raising the question of whether they contributed to a media environment that encouraged and fostered a diversity of viewpoints. We reasoned that if podcasts did indeed offer a voice to a greater variety of individuals, then, perhaps, the diversity of media forms and distribution channels might legitimately be thought to have become more democratic, with a wider range of individuals gaining access in order to disseminate their views. At least in the case of podcasting, we surmised, one inexpensive form of media distribution had circumvented the traditional need for concern about licensing, ownership, and spectrum scarcity.

Matters of definition also concerned us. For example, would "public" media still carry the imprimatur of democratic pluralism, as it was once intended? Would there be cultural interpretations toward the nature of access to channels of information in different cultures? Would podcasting be subject to the same socially practiced norms regarding freedom of speech? And finally, at this stage of podcasting's growth, what types of audiences were willing to go out of their way to listen to podcasts, and for what purpose?

A number of podcasts were examined to develop a framework for analysis. We then decided to consider the following issues as polar extremes: whether the podcast encouraged engagement with ideas, or whether the podcast was indulgent of the podcaster's personal views; whether the content of the podcast could be considered to represent pluralistic views, or whether it was represented as a matter of the podcaster's own opinion; and whether the design of the podcast was intended to serve many or few. Our initial intention was to investigate whether the podcasts that were controlled by individuals (one or a few), unrestricted by corporate or other sponsorship, would indeed provide fresh insights and encourage open discussion and engagement with the podcasters—whatever their point of view.

We approached this study with an eye toward exploration. We were not intending to conduct an exhaustive content analysis, or even to develop a critique of podcasting as a challenge or alternative to mainstream media (radio in particular), though podcasting has been described as "a glimpse into the future of broadcast news" (Potter, 2006). What we discovered, though, is that while podcasting has opened the possibility for dialogue in some cases, the production of different genres of podcasting actually results in a number of different types of public discourse. While the fragmentation of the audience for mainstream media has been of concern to legacy broadcasters, and the continued niche markets have complicated the teaching of "mass" media to students, podcasting appears to be following in the footsteps of mainstream radio broadcasting at this time.

There is one caveat. Attempting to describe what is happening today in podcasting is like trying to capture a moving target. While podcasting is relatively new, the number of podcasters and the number of people who listen to podcasts are multiplying daily. Our results, then, attempt to capture the representative sample of podcasts at a certain time in history. We hope that the research questions we have investigated contribute to the evolving and emerging dialogue on the nature of digital media in an increasingly segmented, mobile society.

A representative sample of podcasts was chosen from one of the several podcast directories available on the Internet (www.podcastdirectory.com), and the contents of three different genres were charted. The first genre included those labeled as "indie" podcasts, a group that typically has a mission of representing different viewpoints than mainstream media within a broadcast range, and are classified by Podcastdirectory.com as "community" podcasts. We also examined podcasts that were highly personal in nature and, we assumed, would most clearly fit into a category of public access media. Finally, we explored a handful of "specialty" podcasts, which ordinarily limit their content to a particular subject (for example, health, technology, science).

Podcasts that were aligned with mainstream media, such as those produced by a network or social agency, were not investigated, because we viewed these as primarily rebroadcasts that served listeners either out of the broadcast range or that sought merely to increase the availability of content to individuals through the Internet, rather than those who might be listening in real time. Finally, the content of the most heavily accessed podcasts for the month of August 2006 were examined to see if the podcast audience demonstrated any predilection for either the podcasts we analyzed or for any other genres that might be classified as "public" or "democratic" in content.

This chapter investigates the content of a small number of podcasts that are produced by individuals and small groups—and not by the mainstream media—in order to understand whether those individuals and groups who exercise podcasting in today's digital world actually return to the idea that every person has the right to make his or her opinions known, and that those voices contribute to community building or, at least, to information sharing. We hoped to find a potential for greater civic involvement through the topics and discussions in at least some of those podcasts. We approached this study with the idea that the diffusion of digital technologies now contributes to making "public" broadcasting possible, and that podcasting, which has been called a form of "civic journalism," would help fill the gap in media forms that contribute to democratic practice and socially involved media participants.

This material is analyzed with reference to the original intentions of public broadcasting in the United States and Canada. These two countries have distinctly different visions of what public broadcasting is about, but both serve the same purpose—to meet the needs of those unserved by mainstream media. Both public broadcasting systems have, as part of their mandate, a public access component. Our purpose, then, is to investigate the content of an emerging technological use to see if the podcasters either mimic mainstream media or exercise what public policies define as public access and public accountability. Before discussing our observations, however, the newness of the medium requires some description.

A Brief History of a New Technology

When podcasting became available in 2000, the optimistic outlook was that, once again, a technology has been developed that could literally allow any citizen to become a broadcaster. Disc jockey Adam Curry (2002) of MTV became the first public figure to use the podcasting service developed by Dave Winer, a software developer. Podcasting does not require a license because it uses online systems to transfer information and there is no use of the electromagnetic spectrum. Through an aggregator, which functions like an organizer of audio information, the message is packaged for delivery to any technology that uses MP3 compression (the most commonly used is an iPod). The transferring of media files is relatively efficient and inexpensive. The result is a subscription service that regularly downloads audio information, though video podcasts are now possible due to downloading technologies like the iPod Nano, which has increased capacity capable of compressing video as well as audio signals.

The term *podcasting* has a few possible origins including using a small pod-like technology to record a podcast, or thinking of the term as "portable online delivery," the most typical definition reflects the type of technology necessary to receive a podcast. Because iPods were some of the first technologies available for downloading information, the "pod" designation from "iPod" makes sense. The term was suggested by Ben Hammersley (2004) in the British newspaper the *Guardian* on February 12, 2004. By October of that year a number of detailed articles explaining how to podcast began to appear online, and in November 2004 the first podcast service provider offering storage, bandwidth, and creation tools became available. Since that time, a number of services and systems have become available, many of which are free and downloadable from the Internet. In 2005, *podcast* was the word of the year as named by the *New Oxford American Dictionary* (Palser, 2006). Since then, a number of other pod-related terms have entered our vocabulary.

Podcasting also brings the baggage of mainstream radio content to the media landscape. Traditional conflict over the ownership of music has plagued more powerful podcasters, but there are now many recordings available to a wider audience designated as "podsafe." In 2005, the term *podmercial* (that is, pod commercial) was used by a radio broadcaster in Las Vegas who also distributed the station's signal through a podcast. As might be expected, it didn't take long for podnography to become one of the offerings, and adult podcasts (also known as sexcasts) quickly followed. A Google search for "adult podcast" showed that on a specific date in 2005, 6,800 were available (MacMillan, 2005).

Traditional broadcasters have readily accepted podcasting as an additional distribution medium. *Fortune*, the *Washington Post*, the *Economist*, the *New York Times*, C-SPAN, and the Fox television network are just a few of the early adopters of podcasting, and the White House sends podcasts in different languages. However, despite this alternative delivery system for mainstream media and those with special interests, celebrity podcasts seem to get the most consistent number of downloads. *The Guinness Book of Records* notes that the most popular podcast in 2006 was *The Ricky Gervais Show*. In July 2005, the first People's Choice podcast awards were held during Podcast Expo, where awards were presented in twenty categories (http://www.portablemediaexpo.com/). It is likely that the primary reason so many mainstream and special-interest broadcasters also podcast is the hope that, while the podcast audiences may be small, they actually provide excellent niche audiences for advertisers.

It is difficult to accurately assess how many podcasts there are, and what the listenership/user statistics are. Bridge Ratings estimates that by the end of 2005 about 5 million people had heard at least one podcast, though a much smaller number (about 1.6 percent of the population) actually podcast themselves (Palser, 2006, p. 65). The most dramatic growth of the new form accompanied the increase in iPod sales throughout 2005 (about 6 million per quarter, each quarter of 2005), which fueled a rapid growth in the number of podcasts, from about 1,000 to 26,000 feeds in 2005 (Bullis, 2005, p. 30). A quick look at Podcastdirectory.com shows hundreds of genres and thousands of "programs"—though there is no way to know how often individual podcasters upload programs. It is the nature of this technology that there are no reliable numbers to indicate how many of those podcasts were actually downloaded or listened to; because podcasts come to listeners through subscription services as well as through individual downloads, there is no accurate measure of whether the downloads are actually heard (Palser, 2006, p. 65), but Bridge estimates that by 2010 there will be a potential audience of at least 45 million (Potter, 2006, p. 64). With this potential audience in mind, mainstream broadcasters and other media outlets have taken it upon themselves to get into the business sooner, rather than later.

Defining "Public"

Though the broadcasting infrastructures in both the United States and Canada had the mission of serving their respective publics at the time of their inception, the concepts of how the public might be served have undergone many changes over the years. Ironically, the Canadian Broadcasting Corporation (CBC) and the system designated as "public broadcasting" in the United States, the Corporation for Public Broadcasting (CPB), were both formed in response to the commercial system of broadcasting that had emerged in the early days of radio in the United States.

To the casual observer, the public broadcasting systems of the United States and Canada may appear not at all dissimilar. After all, both were ostensibly born during a period defined by the post–World War II economic expansion in the West and the proliferation of geopolitical tensions generated by the Cold War. Furthermore, the relatively young systems matured at a time when the citizens of both nations possessed a heightened appreciation for the development of the social welfare state (as evidenced in the United States by the popularity of the Kennedy and Johnson administrations, and in Canada by the majority governments of Lester Bowles Pearson and Pierre Elliott Trudeau) and understandable concern about the growing conflict in Vietnam. Today, both nations enjoy relatively high levels of economic prosperity, live within a sociopolitical environment defined by extraordinary democratic freedoms (if curtailed slightly by new restrictions brought on by the post-9/11 environment), and adhere to cultural norms that are increasingly dictated by consumerist tendencies.

The American experiment in public broadcasting began principally with the 1967 report of the Carnegie Commission, entitled *Public Television: A Program for Action*, which resulted in the subsequent passage of the Public Broadcasting Act of 1967. In its introduction, the report suggests that public television should advance and promote citizen interests that are not covered by commercial broadcasting; deepen a sense of community life; bring to life the richness of policy debate and political discussion; reflect the accomplishments, hopes, fears, and agitations of citizens; and provide a voice for underrepresented groups.

Public radio was not even mentioned in the Carnegie report, but was "injected" into the draft just prior to submission to Congress (Ledbetter, 1997, p. 117), and National Public Radio (NPR) was formed in 1970. The mission statement pledged that NPR would

> serve the individual: it will promote personal growth; it will regard the individual differences among men with respect and joy rather than with derision and hate; it will celebrate the human experience as infinitely varied rather than vacuous and banal; it will encourage a sense of active, constructive participation rather than apathetic helplessness. . . . The programs will enable the individual to better understand himself, his

> government, his institutions and his natural and social environment so that he can intelligently participate in effecting the process of change. (Ledbetter, 1997, p. 117)

Since its inception, however, the noble mission of public broadcasting has been no match for "the wishes of society's most powerful elements—the federal government, the military, and large corporations" (Ledbetter, 1997, p. 36). Given little or no insulation from political gamesmanship, the public broadcasting system (PBS) in the United States has relied heavily upon inconsistent government appropriations (spanning no more than three years but often covering a single year) and varying levels of direct "member" support, along with corporate underwriting that increasingly mimics traditional commercial advertising.

While PBS is one of seven terrestrial television networks in the United States (and is also—by far—the least watched), and NPR has an audience profile that reflects a more upscale, educated listener, CBC competes with only two other national networks and enjoys a much higher market share. Like PBS, CBC functions as a nonprofit corporation. Unlike PBS, however, CBC's mission is more clearly defined as an overt instrument of national cultural promotion (see, for example, Peers, 1979). As stated in the most recent public broadcasting legislation, the Canadian Broadcasting Act of 1991, CBC must "provide television services incorporating a wide variety of programming that informs, enlightens, and entertains . . . must be predominantly and distinctively Canadian . . . and contribute to a shared national consciousness and identity." The CBC continues to provide not only the largest share of original programming in Canada, but nearly all of the nationally distributed programs featuring Canadian creative and performing talent.

In contrast to the American television spectrum, however, Canada maintains a far blurrier distinction between the public and private components. Though modeled largely after the British Broadcasting Corporation (BBC), in which a government-chartered public corporation provides a public service, CBC is also a system based upon an uneasy compromise between public and private interests. CBC affiliates carry commercially driven programming in addition to more traditional public programming, and in doing so take in funds that allow them to show national content and pursue the public mission.

What both mission statements indicate is that the concept of public broadcasting in both countries actively seeks to enhance the democratic process by representing a range of individuals, viewpoints, and cultural values. For the purposes of this analysis, we draw from the mission statements of the PBS and CBC and define "public" as a system that serves a community or region, reflects the values and attitudes of a plurality, and is not commercially funded.

It is difficult to find fault with the desire to serve the public, but according to data on both sides of the border there is citizen dissatisfaction with both systems' capacities to live up to their original promise, with one exception—the niche audience who comprise loyal NPR listeners. Still, these listeners now receive content that does not live up to the original mission of NPR. As R. A. Goodman (1992) has written, the financial aspect of running public radio stations has driven out the quest for directly serving the community and creating content that primarily serves the upscale listeners who listen most often to programs like *Morning Edition* and *Talk of the Nation*, which are some of the most expensive NPR offerings. While a comparison of recent annual reports suggests that the Canadian PBS continues to receive up to ten times the per-capita public support of its American counterpart, Canadians appear no more interested in its content—and, equally important, no more politically or civically engaged as a result of their exposure to that content—than Americans. Could podcasting offer an alternative to these public systems?

A clue to a possible answer to this question comes from the remarkable popularity of the American NPR podcasts. In the United States, NPR and member stations offer almost two hundred different podcasts. As Barb Palser (2006) writes, "The marriage between public radio and podcasting couldn't have been scripted more perfectly: Much of NPR's content is essentially ready-made for podcasting, and listeners were literally begging for podcast versions of shows months before they were available. Now NPR's podcast-only content, with podcast underwriting, provides a new revenue source for both NPR and local public radio stations" (p. 65). It seemed logical to us that with a highly selective audience seeking out one type of public radio broadcasting, there may be a ready-made audience waiting for the type of content U.S. public radio furnishes. With that in mind, we began our exploration.

Listening to Podcasts

As previously mentioned, three genres of podcasts were chosen as representative of what we thought would include fresh viewpoints and cultural values, and that would encourage democratic participation through dialogue and sharing of ideas. What we found was actually depressing.

Podcasts were initially sampled from the classification of "community" podcasts, and the content was charted on the grid of extremes dealing with polarities of democratic values. A representative sample of podcasts created by individuals was then chosen and coded. Last, a range of specialty podcasts whose respective creators necessarily limited themselves to predetermined topics were examined. The skew of content is represented in table 1.

Table 1: Grouping the Three Genres of Podcasts According to Their Polar Extremes

Engagement	**Pluralist**	**Serves Many**
Community podcasts	Community podcasts	Community podcasts
Specialty podcasts	Specialty podcasts	Specialty podcasts
		Specialty podcasts
Individual podcasts	Individual podcasts	Individual podcasts
Indulgent	**Personal**	**Serves Few**

While podcasts in these genres clearly fit our definition of "public" radio as locally based media that is noncommercial, the range of viewpoints, ideas, and values were widely divergent. Even though the production values of the podcasts were, for the most part, as good as those of mainstream radio, and often more creative, the content of the community and individual genres skewed toward the "indulgent" and "opinion" categories of our grid. The extremes were so great that we felt we needed to consider how podcasting was either similar to, or different from, mainstream radio, and then to examine how podcasting was unique.

While the podcasts did indeed bring a sense of localism to the audience, they were highly targeted to niche audiences. There were extreme differences between the community podcasts and those sponsored by individuals, as might be expected. While community podcasts tended to represent a form of radio that is rapidly disappearing in the United States and Canada because of chain ownership, rarely was there an approach to content that would inform (except about an event) or explain any contemporary issues. Podcasts by individuals actually represented ham radio content that was very personal in nature, though they were marked by expletives and humor that could be classified as raunchy or scatological (see table 2).

Table 2: Podcasts Representing Different Genres

Community Podcasts	Individually Produced Podcasts	Specialty Podcasts	Most Heavily Accessed Podcasts, August 2006
Orlando Events	Bibb and Yaz	Brain Food	Sex with Emily
The Perfect Face	Acts of Volition	Cat Lover's Podcast	Rumor Girls
Bad Cop, No Donut	Frickin Circus	The Multiple Sclerosis Podcast	Preying Lizard
Smart City	Hometown Tales	Winecast	Cute & Sexy
Kabbalah Forum	3 Guys, 3 Beers	SCUBA Chat	Daily Source Code
BuzzNet News	The Ammonia	Around My Kitchen	World Cup 2006
JUB Radio	The Jackie Radio Show	Baseball Historian	ABC News Extreme
Luna Musings	Nontourage	Sound Medicine	Evil Genius Chronicles
AuRanach Deorar	2 Rednecks	Health & Beauty Live	IT Conversations
Nuclear Winter	The Mike Show	Green Livin'	Ask a Ninja

The community podcasts did indeed include more locally based information about events, but for the most part, the podcasters featured on the program overwhelmingly seemed to treat the podcast like a radio program that focused on arts and activities in the region. Often different segments would be hosted by individuals who would interview community members about their personal approaches to whatever they did as artists, citizens, or community activists, but little debate or discussion of ideas actually occurred. After listening to a number of these podcasts, we began to feel that these (though only a small sample of podcasts available) were informative about, though not necessarily representative of, issues within a community. Surprisingly, even though no commercial sponsorship was identified, U.S. podcasts tended to have a sizable number of commentaries on local eateries and business establishments in the area, reminiscent of sponsorship.

Opportunities for dialogue in the "community podcasts" were few, with very little encouragement for podcast listeners to let the hosts of programs know what they thought of the program, or to give feedback or comment. Noticeably, most of these podcasts did not even offer an email address or have a Web site, which might, given the medium, seem like a logical way to encourage feedback.

The individual podcasts were very different, however. It almost became difficult to listen to many of these, because they reflected such highly personal commentary about, for example, the sexual habits of the hosts and who they thought was "hot" in a sexual sense, and the podcasts almost appeared to be little more than an opportunity to use coarse language and put-downs. The sense of personal indulgence in the individually oriented podcasts seemed to flaunt both indecency and immorality—in every justifiable sense of the word.

While these podcasts raise the question of whether individuals can and should become "citizen journalists," one has to wonder who would want to listen to such personal indulgence. We know that there are wide ranges of personal taste and humor, and that some individuals enjoy this type of commentary, but the constant harangue of the content leads one to question how much time any podcast listener could or would spend consuming these rants. These podcasts in particular raised questions about freedom of speech and indecency, and while we advocate all aspects of freedom of speech in the media, we note that the content of individual podcasts almost beg for greater discourse surrounding appropriate content and indecency in podcasting. Analyzed from the perspective of cultural critics, these podcasts may well indicate the predominance of changing social mores and norms with regard to social speech.

Of the three genres examined, the group of specialty podcasts afforded us the greatest optimism that the new medium was, at least in part, advancing the public mission. Nearly all of the podcasters seemed to relish conversation with listeners. The majority offered their personal email addresses or other points of contact, and many made time to read electronic feedback or included listeners in later broadcasts. Though some were deeply personal (for example, one couple's confrontation with multiple sclerosis; a single mother's quest to offer her four children nutritious meals) and others more lighthearted (for example, a review of dessert wines or recommendations for good scuba diving spots in the Caribbean), all functioned to offer some degree of prescriptive value to the listening public. Furthermore, as they all lacked commercial backing, sponsorship, or endorsement, the individual podcasters were clearly motivated by a simpler desire to share something they enjoyed with others who possess similar interests. Consider, for example, how the producer of *Brain Food* explains the impetus behind his podcast:

> I am an undergraduate student studying Computer Engineering in Ontario, Canada. Throughout my academic career, I have always felt a passion for teaching and explaining topics to friends, family, and peers. My new project, the Brain Food Podcast, seeks to be an extension of that passion. With my background in Science, Mathematics, Physics, and Chemistry, I seek to take everyday mysteries and explain them using powerful analogies and simple logic in a way that everyone scientific and non-scientific alike can appreciate. (http://www.brainfoodpodcast.com)

If the specialty podcast genre gives us reason for hope, however, recent rankings of podcast popularity are far more sobering vis-à-vis the ability of podcasts to serve as a new form of public media. As we considered the range of what appeared to be the most popular, based on the number of worldwide downloads, we had to remind ourselves that the number of downloads does not necessarily equal listenership, because of the subscription feature of MP3 automatic downloads, and, at this point in history, the population of podcast listeners tends to skew toward the younger demographic.

Among the top ten most downloaded podcasts for August 2006 (see table 2), three focused primarily on sex, one on popular music, and two on world events. Only one of the podcasts, *Ask a Ninja*, seemed to focus on the personality of the podcaster and, therefore, resembled mainstream radio far more than the other categories represented in this group.

Generalizations

While this exploratory study randomly sampled a variety of podcasts from three genres—community, individual, and specialty—and briefly summarized the content of the most downloaded podcasts for one month, we feel confident in reexamining our original research questions that focus on the topic of whether podcasting offers an alternative to public media. While it would be wonderful to be able to give a simple yes or no answer, as with most questions in life the answer is more complicated than that. What we have concluded about podcasting, however, may inform other researchers interested in pursuing this topic, and, because podcasting is constantly evolving, this snapshot in time grounds future research in some ways.

Perhaps the most noticeable feature of podcasting as it relates to mainstream radio is that the niche audience is the key audience served for both types of content. This leads us to ask the question of whether radio, as a medium, has changed so dramatically from its original sense of both uniting an audience and providing local content that it has become something other than that for which it was originally intended.

If this is so, the future of public media is on a very unfortunate trajectory. Pluralism, democratic debate, and a sharing of viewpoints may exist on specific podcasts—particularly those involving specialty topics—but this means that the characteristics of niche audiences and certain genres of podcasting do not live up to the traditional sense of what public media may have meant at the time of the establishment of the two major institutions of broadcasting in the United States and Canada. This is not necessarily bad, but it does show a shift in the values represented by forms of media.

Podcasts may be an expression of freedom of speech, but the genres of podcasts that we explored are not necessarily representative of a public space for the actions of democracy. As a marker of our cultures, these podcasts provide information, but not always diverse viewpoints or values that inform. They would most closely fall into the entertainment function of media, and perhaps that is what they can do best.

What we were not able to observe from this exploration into what could be called "small-format" media is whether the audience or potential audience actually feels that podcasts legitimate civic discourse or civic involvement. Erik P. Bucy and Kimberly S. Gregson (2001) analyze a number of media formats, including both traditional and "new" media, to examine the distinguishing features of new formats and the potential empowerment aspects contained in new media forms for civic engagement. They too find not only that different formats appeal to different segments of the audience, but also that one of the most important features of any new form of media is the contribution it can make to the "agenda-building process of public issue formation" (p. 376). Moreover, their thesis discusses the role of the stability of civic discourse and political participation, and reminds readers that elites often become the most active in contributing to issue formation. Their work also reminds us that research indicates that mediated political experiences provide psychological rewards and personal empowerment for those who have "civic" media experiences (pp. 369–370).

If we were to address these levels of civic satisfaction and civic involvement that might be available through podcasting, we would need to consider how podcasting contributes to the many forms of engaging in the expression of new ideas, along with the multimedia use of most individuals in North America, and we would need to consider the contrast between knowledge about political issues and actions taken toward creating political change.

Conclusions

Podcasting is still in its infancy, and, as it develops, traditional categories of genres might grow to become more inclusive and more representative of public spaces for actual democratic discourse. Obviously, the next step of our research would be to look at the podcasts sponsored by different political factions and interest groups to see how inclusive they are with regard to the definition we have crafted about what constitutes public media. Without a doubt, the number of podcasts available suggests that there are widely divergent topics available. The ultimate question is whether those who seek these podcasts do so exclusively or whether they expand their own searches to inform themselves about diverse viewpoints.

What we found when exploring the various podcasts with a traditional definition of what "public" media means is that the elements of public access, civic discourse, and political involvement have all changed in nature. Public access to diverse content and ideas may be available, but one must search for the content that will present new ideas and values. With subscription services, like podcasts, the search may not be constant. Once someone finds a podcast that they like, there is the possibility of subscribing and having multiple programs available with the luxury of listening to them in one's own time. The problem with this type of engagement, though, is that while an invitation may be extended to comment, or engage in discourse, the "conversation" will probably not exist in real time. For the most part, any commentary or engagement in dialogue will be extended over time, as someone responds via email. The instantaneous nature of response is highly limited by the technological structure of podcasting.

What has not changed, given the current snapshot of who listens to podcasts, is that, when the broader audience is considered, podcasts that primarily serve entertainment purposes overwhelm those that encourage participation. The success of NPR podcasts with an upscale, elite audience is reassuring, especially for those people who consider themselves to be politically aware, but the range of topics is necessarily broad and avoids any local community connections.

Where we hoped podcasts would be most important—at the community level—was the place where we were most disappointed with the ability of podcasting to offer a public place for discussion. Whether these podcasts represent local radio that is more concerned with providing information and events listings rather than contributing to participatory dialogue is a matter of podcasting's youth, or whether the audience for podcasting is currently too small to explore the potential for involving members of the community in civic discourse, the local community is still, at this time, underserved.

Further studies of the audience of podcasts will also undoubtedly present snapshots in time for the evolution of podcasting, and, as for all media, a far richer pool of information would be gleaned if a more rigorous study of audience preference and use could shed light on how audiences use this form of communication. With many more technologies and distribution systems vying for audience attention, and individuals having to make choices about how to spend money and time, there will be fewer opportunities to link any one form of media with explanations of how it discretely contributes to individuals' sense of self and participation in society.

What this study does, though, is explore some genres that are represented in podcasting during the medium's infancy. Will podcasting move from a position of mimicking mainstream radio to become something unique in and of itself? We revel in the possibility for Internet distribution of a wider variety of content, but at this point the content of local, community podcasts and those podcasts that are produced by "citizen journalists" at a very personal level lead us to think that as these channels of communication become available they will continue to have more than their fair share of fluff, indulgent opinion, and sexual content as well as profanity. As tests of democracy, they suggest that the new technology is worthy of study, but as a form of public space for debate they lead us to conclude that the new technology is very much like the old technology.

References

Bagdikian, B. (2004). *The new media monopoly*. Boston, MA: Beacon Press.

Bucy, E. P., & Gregson, K. S. (2001). Media participation. *New Media & Society*, 3, 357–380.

Bullis, K. (2005, October). Podcasting takes off. *Technology Review*, 30.

Carnegie Commission on Educational Television. (1967). *Public television: A program for action*. New York: Bantam Books.

Curry, A. (2002, October 21). Cool to hear my own audio-blog. Retrieved from http://radio.weblogs.com/0001014/200310/12.html#a4604.

Fowler, M. S., & Brenner, D. L. (1988). A marketplace approach to broadcast regulation. *Texas Law Review*, 60, 209–257.

Goodman, R. A. (1992, Winter). Uneasy listening. *Whole Earth Review*, 105–109.

Hammersley, B. (2004, December 2). Audible revolution. *Guardian*. Retrieved from http://www.guardian.co.ui/online/story/0,3605,1145689,00.html.

Ledbetter, J. (1997). *Made possible by . . . : The death of public broadcasting in the United States*. New York: Verso.

MacMillan, R. (2005, August 11). The paradox of podcasting. *Washington Post*. Retrieved October 11, 2006, from http://www.washingtonpost.com/wp-dyn/content/article/2005/08/11/AR2005081100695.html.

McChesney, R., & Nichols, J. (2002). *Our media, not theirs: The democratic struggle against corporate media*. New York: Seven Stories.

Palser, B. (2006, February/March). Hype or the real deal? *American Journalism Review*, 65.

Peers, F. W. (1979). *The public eye: Television and the politics of Canadian broadcasting, 1952–1968*. Toronto: University of Toronto Press.

Potter, D. (2006, February/March). iPod, you pod, we all pod. *American Journalism Review*, 64.

· 9 ·

REACH OUT AND DOWNLOAD SOMETHING

An Analysis of Cell Phone and Cell Phone Plan Advertisements

Richard Olsen

> *Machines, images of machines, discourse about machines, and aesthetic dimensions of machines have long been understood to contain potential rhetorical power. . . . The machine as object, image, or subject of discourse is used to manage meanings and commitments in a society.*
>
> Barry Brummett

In his groundbreaking classic, *Diffusion of Innovations*, Everett Rogers (1995) discusses five types of people and the way they respond differently to new ideas and technology. The innovators are most eager to adopt new things—perhaps simply because they are new. They are followed in order by the early adopters, the early majority, the later majority, and finally the laggards, who share the Luddites' general mistrust of most things new. The cell phone has penetrated well into the later majority category of users, and it seems now almost an act of cultural resistance not to own some form of cell phone. Those who have cell phones will speak to the "necessity" of the device, but how did such "necessity" get established? Rogers explores, among other things, the roles that various forms of communication play in fostering the adoption of new technology: word of mouth, local opinion leaders, media campaigns, and the like can all play a part. While exploring each possibility could prove interesting, this chapter focuses on how advertisers influence and reflect the expectations and common understanding of cell phone performance and cell phone plans. The goal of this exploration is to help explain what cell phones mean and to reveal what they symbolize within American culture.

The ubiquity of cell phones is perhaps exceeded only by the proliferation of cell phone stores and subsequent advertising to get people into these stores. According to *Advertising Age*, Verizon, Cingular, and Sprint spent a combined $4 billion on media advertising in 2005 (Shumann, 2006). It is beyond the scope of this chapter to examine all the strategies employed by these and other phone companies in their advertising. However, a convenience sample of several hundred ads and plan brochures was collected by students in several communication studies classes and used as the primary sample for this chapter. This sample was used as an illustrative set of artifacts, rather than as an empirically generalizable sample.

Advertising and Consumption of Technology

In *Machines That Become Us*, James E. Katz (2003) says that the purpose of the collection of essays is to explore "the relation between the externally created machine and internally created reality" (p. 18). That is a concern of this essay as well. Advertisements play a critical role in that process by offering implicit and explicit arguments for how we should understand and use things.

It is difficult to determine advertising's unique contribution to an outcome, as Kim Rotzoll and James Haefner (1996) have noted. The complexity of culture and human nature make it impossible to isolate a specific cause-effect chain. Leslie Haddon (2003) echoes this conclusion, noting that "how one experiences ICTs [information and communication technologies] is not completely predetermined by technological functionality or public representations; it is also structured by the social context into which it is received" (p. 147). In light of such insights, I am not arguing that the content of the ads examined fully explains what cell phones mean or why they are used the way they are.

However, it is true that advertising manipulates symbols to create meanings, and often those meanings and values mirror the dominant ideological themes of the culture (Frith, 1997). Rotzoll and Haefner (1996) note that advertising has four fundamental goals: (1) precipitation or creating awareness and need, (2) persuasion to purchase through emotional and rational appeals, (3) purchase reinforcement, and (4) reminders that reinforce brand loyalty (p. 117). Notice that in goals 3 and 4 there is not only the attempt to reinforce specific purchases but to reinforce the act of purchasing in general. This is a theme that will be explored later in this chapter.

A final complication to analyzing ads regarding cell phones is that we live in a society of ads and layers of culture that rely on one another: "We are inundated with images and signs that no longer have referential value, but instead, act solely with other signs. This marks the advent of simulation. Instead of the previous vertical connections between signifier and signed, there is now only the horizontal relationship of signifier to signifier" (Kraus & Auer, 2000, p. 2).

The following assumptions and tools are used to help bring some clarity to this messy stew of texts and culture.

Of Method and Mosaics

Barry Brummett (1991) suggests that the "critic's business is to show how patterns available in a cultural repertoire could have been used to order the symbolic environment within a rhetorical transaction" (p. 95). The goal is to illuminate and evaluate patterns of persuasive texts of which cell phone ads and documentation certainly qualify. The intertextual relationships among such texts (Fiske, 1989) provide a rich and multilayered artifact to unravel.

Two key factors delightfully complicate the analytical project at hand. One is that metaphor—broadly understood as the "thisness of that" by Kenneth Burke (1969, p. 503)—runs throughout cell phone texts. Indeed the cell phone itself can be a type of metaphor for other things. Second, many of the texts examined in this chapter are highly visual in nature. To help make sense of such features, the work of Paul Messaris (1997) is drawn upon to help define visual metaphors and visual propositions. Dann Pierce's (2003) typology of visual metaphors that parallels traditional discursive categories that includes metonymy (association), synecdoche (part for whole), and replacement (substitution) is also invoked.

While metaphor and attention to visual configuration will be addressed, the analysis is structured more explicitly around major fantasy themes. Fantasy theme analysis examines how messages shape and reflect social reality and provide communication artifacts (jokes, vocabulary, stories, T-shirts) that foster symbolic convergence (Bormann, 1982). A successful ad, for Earnest

Bormann, will reflect back to its viewers a fantasy that dramatizes the real or the ideal of that viewer. If the ad is successful in doing that, it will foster a sense of convergence or unity around that symbolic content. Smaller dramas and storylines are referred to as fantasy themes. An overarching drama or narrative that expresses the ideals and goals of the group is known as a rhetorical vision (Bormann, 1972).

This approach is particularly well suited for advertising because, as Roderick Hart and Suzanne Daughton (2005) note, advertising focuses on ideals: "contemporary ads emphasize themes of self-transformation—they do not reflect what people are doing but what they are dreaming" (p. 198). While the implied drama may be as much about dreamy ideals as factual reality, when conducting analyses one still asks questions about the settings, personae, real or implied plots and actions, dominant emotions, and the like. The goal, consistent with Brummett's point above, is to discern and evaluate patterns among the mosaic that likely inform the rhetorical transaction, formation, and sustenance of social reality.

There are a myriad of potential themes to explore within cell phone ads. Certainly there are traditional fantasies of sex, beauty, and wealth. But such analysis doesn't get us very far: cell phone ads often use strategies common to other forms of advertising. Rather, in this chapter, the findings have been organized based on a typology similar to Abraham Maslow's hierarchy of needs (Borchers, 2005). This scheme works well because it helps illuminate how cell phone advertisers and plan promoters position the cell phone at one point or another to help meet all of our basic needs. More important, it builds toward the dominant rhetorical vision that frames cell phone advertising: the cell phone as necessary technology for the transcendent consumer. But before focusing on this, it's important to briefly address some minor themes in the ads.

Some Surprising Minor Themes

Many of the ads that were examined contained more than one theme. So saying that quality, price, and technological features were "minor" themes does not mean that they were not present that much in the sample analyzed. Rather, they rarely were the dominant theme presented in any given ad. Of the 312 ads collected, only 6 had a dominant appeal to quality. A Verizon ad claiming "Best in Wireless," because their customers deserved it, that went on to outline how they were "The Leader" in customer loyalty, reliability, and other key criteria, was the clearest representation of this small category. For the vast majority of ads, quality was implicit in the metaphors or fantasy themes used in the ad or a subordinate theme to a more dominant claim or narrative.

Technology was treated in a similar way. While every ad at some level was about the technology, few ads made it the primary focus. The Verizon slogan "It's the Network" could of course be interpreted as being about the quality of the company or the unique technology of the company. Ads for Treo phones, BlackBerry devices, and a Nokia phone with replaceable multimedia cards were typical of the ads that focused on technological advances.

It was far more common to focus on the products made available through technology (music downloads) or the social results of technology (capturing the picture), rather than focusing on the technological potential directly. In this sense, technology was part of a larger fantasy theme or narrative of better living through technology. Technology is paradoxically central and subordinate within this fantasy.

Nature and anthropomorphizing of the phone were also minor but important themes. Within the fantasies examined later in this chapter, the phone is an implied character in the drama. Making the party happen, technology is a "hero" of sorts. However, few ads openly framed the phone this way. A Motorola ad described its wireless earpiece as "Beautiful. Smart. And Surprisingly Unattached," framing the phone as an attractive significant other. Virgin Mobile used a similar strategy with a brochure that asked if the user was "Looking for a new kind of relationship." Framing the cell phone as "natural" was very rare and seemed to be a primary focus only with the Pebble phone used by T-Mobile: "Looks like a pebble. Feels like a pebble. But twitch your hand just slightly and this natural clam opens smoothly to reveal 21st century technology." Notice that this ad also blends nature with an explicit appeal to technology, illustrating the interplay between these themes. The image accompanying the ad is a very contemporary "Lady of the Lake" female with white sand, blue water, and green hills framing her and the phone.

A Hierarchy of Appeals in Cell Phone Ads

Advertising Hall of Fame inductee (1998), Marion Harper, Jr., once noted that "advertising is found in societies which have passed the point of satisfying the basic animal needs" (AAF, 2006). Yet those basic needs can still appear in ads. This next section moves through a modified version of Maslow's hierarchy of needs because, in general, ads appealing to higher levels were more common than those appealing to more basic needs and because it helps illustrate how the cell phone is positioned to meet "all" our needs. The levels focused on are: (1) safety and security, (2) belongingness, (3) esteem and empowerment, (4) self-expression, and (5) transcendence.

Safety and Security

Ads that appealed to safety and security needs were not as common as they seemed to be in earlier eras of cell phone marketing; the narratives of being able to call when the car won't start in a darkened parking garage, and so forth, were not present in the sample collected. Cingular offered an advertisement based on the idea that "4 of the top 5 commercial banks use Cingular for wireless email" as an appeal to network security, but few other ads focused on encryption issues. Cingular also made safety an issue with their Firefly phone for kids, showing a seven-year-old boy alone at school with the caption "Imagine you're seven. You miss the bus." SunCom also offered a roadside assistance plan to help protect customers from an unpredictable outside world.

Ironically, there were many ads offering customers protection from an unpredictable cell phone industry. SunCom made this a central part of their "Truth in Wireless" campaign that is clearly a play on the "truth in lending" requirements of banks. They suggest that SunCom is combating the confusing contracts that contain "small print or sneaky clauses" other companies villainously offer. Tracfone also frames itself as a safer choice because of the entrapping nature of long-term service contracts. A key subtheme for each of these campaigns was independence, while a more dominant theme in the ads was one of belonging to a community.

Jim Kuypers (2005) notes that in doing fantasy theme analysis it is important to pay attention to dramatized messages and how those contribute to an overall rhetorical vision. With each identified theme, the dominant drama is briefly summarized. Later, these dramas are pulled together to reveal the rhetorical vision they construct for cell phones and their users.

To summarize this category of appeal with respect to key fantasy themes: the setting is either dangerous or ambiguous. The villain is implied—danger could come from this situation. The victim is the cell phone user (or person in need of a cell phone to use). The hero is the technology sponsored in the ad. It is important to note that none of the ads in this sample made a "strong" fear appeal. Even the left-behind child outside the school was treated with humor. We seem to be saving Junior from frustration or low self-esteem, not from potential abduction or assault.

Belongingness

Belongingness was a significant theme within the ads. This should not be surprising given that companies are selling communication technology. This theme is consistent with Nicola Green's (2003) finding that mobile devices are often "talked about in positive and emotional terms as 'connection with (peer) others'" (p. 203). What is interesting is the diversity of this theme. While almost all of the samples collected were for American cell plans and companies, one student made me aware of a campaign for 1NDIA with the slogan "One Nation. One Voice. One Rupee." Verizon launched its "in" campaign to appeal to the struggle of community versus isolation. One ad features a group of friends standing together looking out at the ocean. They have their arms around one another in a clear display of platonic affection. A red circle includes all but the young women on the far right. The text "Are you in?" is above the heads of those in the circle while the text "Or are you out?" is next to the excluded women.

Alltel made prominent use of the notion of a circle of community in its "Circle of Friends" plan. Visually, the theme of belongingness emerges with a group shot of friends that clearly portrays unity, connection, and togetherness. A Motorola ad with the caption "Buddymoto" featured three girls who apparently go to a private school with a dress code (plaid skirts, white shirts, and brown sweaters), which would suggest conformity, yet they are able to express individuality through unique shoes, hairstyles, a sporty cap, and such. Thus they are a unique group and their cell phone is part of that group identity.

In addition to friendship, family was also an important theme within this category. The most obvious pitch was a Cingular ad promising the ability to "Have the family meeting without the family meeting" through their push-to-talk (similar to walkie-talkie) technology. Connecting across generation gaps was also somehow easier with cell phones. Sony advertised a phone by picturing a stereotypically stoic dad and retro-punk daughter next to each other with the caption "If only they'd talk." The ad copy goes on to imply that the advanced microphone qualities of the phone might help them have a heart-to-heart.

An interesting aspect of belonging capitalized on by a few companies is exclusiveness. There were a number of advertisements and services that emphasized belonging to an exclusive group. One example is ring tones that adults find difficult to hear. Mosquitotone is just one name given to ring tones at around 17 kilohertz that younger ears can hear but most adults cannot (Vitello, 2006). In this case, belongingness is defined by who is excluded.

Many services are also promoting a belongingness theme as part of their pitch for subscribers. ESPN, *Playboy*, *People*, *Cosmopolitan*, *For Him* (*FHM*), and *Teen Vogue* are among those that offer membership services that include exclusive text and image downloads via cell phone. For example, the ad for People Mobile stresses the words "Premium Mobile Club" and "Join Now" as part of their pitch. The impact of downloads is addressed later in this chapter. For this section it is important to see that the download providers previously mentioned, and others like them, provide downloads in the context of belonging and membership, not as an individualistic or episodic practice.

To summarize, in terms of fantasy themes, this set of appeals reveals a different configuration than that of safety and security. The setting is far less ambiguous and typically social and nonthreatening. The villain is isolation or exclusion, and the victim is the person who chose the wrong network or didn't subscribe to the cool service. The hero status here is shared between the technology and the person who has made the right choice regarding the technology. The schoolgirls, for example, in addition to fulfilling certain fetish-laden stereotypes, are presented as empowered and not merely rescued, as with the security themes.

Self-Expression

For Maslow, the next need is that of esteem or a sense of worthiness in the eyes of others. Advertisers cannot sell the approval of others, but they can suggest that they sell the stuff required for the self-expression needed to garner the attention and/or approval of others. LiAnne Yu and Tai Hou Tng (2003) are among several scholars who have noted that many teens view the cell phone and other technology as jewelry. Thus it can convey social status and "street cred" (Green, 2003, p. 205). The type of cell phone used is meant to convey or reflect the consumer's personal taste—much like a unique piece of jewelry would.

This perception of technology as self-expression is revealed in the Motorola campaign that links its "MOTO" with various prefixes—for example, "MOODYMOTO" and "GOSSIPMOTO." These ads frame the phone as both enabling and expressing one's core characteristic. LG invoked the common narrative of a woman going to a party and finding someone there wearing the same outfit: "I can't believe she has on the same phone as me." The fine print of the ad promises that the sophisticated design "will get just as much attention as anything else you are wearing."

Image was important in a variety of settings. A Sanyo ad encouraging consumers to "Express Yourself" featured photos of business, formal, and casual settings—each one calling for the use of a different snap-on shell for the phone. A Cingular BlackBerry ad reminded users to "take care of business and look good doing it."

Self-expression was not limited to the look or features of the phone, though these were clearly the dominant appeals in the sample. Custom ringback tones (tones that ring a certain way on other people's phones when the purchaser makes a call) were also a key feature for buyers to consider. A Verizon ad stated, "Because before they hear me, they hear my ringback tones." In the ad four young males are in various states of happiness over recognizing or being recognized by their friends through various ringback tones.

The concept of self-expression within cell phone ads relies on a deeper concept within American advertising: choice. *Choice* is a "god term" in advertising, "that expression about which all other expressions are ranked as subordinate" (Weaver, 1970, p. 88). Contemporary American consumers must have choice. The old Henry Ford line that his customers could have the Model T in any color "so long as it's black" certainly would not work today. One might counter that the popular iPod was first only offered in white, but that gave rise to an entire industry of products to customize the iPod.

Choice shows up in cell phone ads in meaningful ways. There are significant differences among cell phone plans, "unplans" (simple plans that are framed by advertisers and companies as an escape from more traditional plans), and prepaid minutes. There are significant differences among phone features, such as email capabilities and built-in cameras, and phone styles, such as flip phones. However, many of the other choices center on, or are framed as, choices of self-expression. Once again we turn to the MOTO campaign with "MOTOFREEDOM." Verizon's "FreeUp" prepay and "America's Choice" calling plans are examples of payment options that feature choice. An ad announcing the Sprint-Nextel merger told consumers to "Get ready to have more choice" and then showed an elevator panel with various sideways buttons in addition to the traditional up and down buttons.

The varied nature of self-expression makes summarizing this category into a single fantasy theme or narrative challenging. However, the dominant setting is a social scene. The villain in this case is the social faux pas of either blending in or selling out—both negative manifestations of conformity. The hero once again is a combination of the technology and the savvy consumer who has purchased and displayed such technology as means of self-expression.

~~Cell~~ Sell Phones and the Transcendent Consumer: The Rhetorical Vision of Cell Phone Ads

The concept of choice can be empowering to consumers. Competition among service providers and cell phone manufacturers can improve products and features, lower rates, improve services, and so forth. However, choice still ultimately frames human beings as consumers and consumption as an ideal way of life.

The final two stages for Maslow are self-actualization and self-transcendence. Actualization is the concept of becoming more fully "me." This involves a close alignment of action and being. Transcendence refers to then looking beyond oneself to issues of spirituality, altruism, and the like. As laudatory as these states of being may seem, they are largely antithetical to the corporate agenda. But the impulse of advertisers to link their services and products with such aspects of our human nature exists, and they do so with another god term: *more*.

"More" is the intersection between the corporate need to create and sustain consumption and the human impulse toward self-actualization and transcendence. T-Mobile offers an ad of a jogger in an exotic location with the text, "I checked my e-mail after I finished the conference call. I replied to the office back in the States AND we raced, my usual run becoming anything but." Smaller text elaborates with, "Who says you can't have both work AND life?" The emphasis on the word *and* is a key connection with the concept of more. More also ties closely with the theme of choice previously discussed. These themes come together explicitly in a SunCom brochure offering "More choices. More ways to stay connected."

"More" in the context of the ads above is tied to maximizing one's natural potential. The jogger got maximum efficiency and enjoyment out of that moment because he was fortunate enough to have a BlackBerry with him.

There is another sense in which "more" can take us outside of ourselves and appeal to the transcendent. This is not just a matter of advertising themes but of the unique characteristics of electronic media. As Brummett (1999) argues,

> The ability of electrotech to create a cyberspace, an inner depth governed by the imagination and mental capacities of its explorers, makes it especially conducive to the creation of utopian fantasies. Utopias are inherently rhetorical, of course, creating as they do standards against which real societies and actual lives may be compared. A cyberutopia may be especially appealing in that its simulations seem real and thus achievable. (pp. 81–82)

The utopian or transcendent aspects of life with a cell phone manifest in three basic ways. First is the "manipulation" of space and time. T-Mobile offers an ad with a barefoot customer in the passenger seat of a nice car, feet on the dashboard, overlooking a remote lake and distant bluffs. The person is positioned as "above it all," and the caption reads, "Let the office come to you for a change." In this ad the traditional spatial relationships of workspace and leisure space disappear. Power dynamics in a typical office are such that workers can't do things on their terms, but that relationship is inverted in this ad thanks to the cell phone. A Treo phone ad offers similar transcendence of time: "It's time for the future to be on your phone." The rest of the text speaks to the convergence of various technologies occurring on a Treo phone, allowing work to be done more efficiently. The picture on the phone, however, is of one of the user's daughters with the caption, "The girls say hi J." This blending of personal and professional life suggests an ability to be two places at once, which is a unique mastery of both time and space.

The second transcendent theme is a utopian fantasy of the infinite. There is a strong connection between this theme and earlier themes of "choice" and "more." However, both of those can be understood to be choices within a finite quantity. Appeals to the transcendent consumer suggest no such limitations. Verizon lets readers know that Verizon as a provider will "never stop working for you." The company markets its high-end mobile all-in-one units by suggesting, "If you want to do it all, we suggest devices that do the same." The ultimate terms of *never* and *all* move us into the transcendent utopia. Verizon pushes even further with an ad with the headline, "INfinite freedom." SunCom pushes us beyond time and space into the infinite by promising, "Any friend. Any relative. Any person. Any carrier. Anytime." The repetition of the word *any* moves the reader into the fantasy of the unlimited.

Each of the preceding themes moves us closer to this final transcendent theme: I can create a world in which I am the center. Life comes to me on my terms.

A mild version of this theme can be seen in a Verizon ad that shows the user at a busy park. While others sunbathe, bike, play sports, and talk, this user watches a download of a Green Day music video. The caption reads, "Out-of-Home Entertainment System." This is a manipulation of time and space because this is typically an in-home activity and the download happens when the user wants, but it is also an appeal to being at the center. Sprint pushes this theme a bit further with ads that ask us to "imagine a phone that turns the world into your living room." This is not simply the inverse of the Verizon pitch but goes further by implying that the entire world, not just entertaining downloads of television shows, movies, and music videos, is available on our terms in our space. Cingular seems to promise a greater sense of dominion with its brochure that asks us to "Call the world your own."

Trends in the content of contemporary advertising certainly contribute to the presence of such themes in cell phone plans and advertisements. Scholars have noted that the rhetoric of contemporary advertising focuses on creating perpetual anxiety and discontent (Hart & Daughton, 2005). Creating a fantasy around the infinite and the ultimate certainly encourages such outcomes. There are also technological reasons for such themes.

As noted previously, cyberspace and electronic media more generally are conducive to utopian and transcendent themes. The popular film *The Matrix* is one example of this potential. Brummet (1999) invokes David L. Altheide in noting that (for those of us not named Neo) the keyboard has become the dominant interface with this electronic world (p. 59). Important to our discussion is that the cell phone either indirectly (with each number having several letters associated with it) or directly (in devices such as the BlackBerry) is a keyboard and has recently been harnessed as such to more fully connect with cyberspace and all that is available through this connection.

The key implication of all of this is that the cell phone is now being defined as a download device, not an outreach device. It is a device to usher in a consumer utopia, not merely to sustain relationships. This "advancement" is hardly without tradeoffs. While the user has instant access to merchants, there is also a sense that merchants are constantly in touch with consumers. Social critic Douglas Rushkoff (1994) illuminates the complex power relations at stake:

> The only place left for our civilization to expand—our only real frontier—is the ether itself: the media. As a result, power today has little to do with how much property a person owns or commands; it is instead determined by how many minutes of prime-time television or pages of news media attention she can access or occupy. The ever-

> expanding media has become a true region—a place as real and seemingly open as the globe was five hundred years ago. (p. 4)

The power of the cell is now not only the ability to reach and be reached by friends and family but to demonstrate "ownership" of media product.

The scene for this transcendent consumer narrative takes place anywhere—revealing the tie to the manipulation of time and space theme. The villain here is not isolation but boredom. The primary concern of teen culture is to avoid boredom (Jagodzinski, 2004, p. 183). This is seen in the earlier Verizon ad that framed the cell phone as a portable entertainment device even though the user was surrounded by a park filled with people—a reality that seemed potentially stimulating enough. It is also seen in a ringback tone ad that encourages downloads so that friends will "hear a song instead of that boring ring." With the increasing emphasis on a 24/7 culture, the mission to avoid boredom no longer seems unique to teens. Notice that in this narrative theme one is encouraged to avoid sensory deprivation, not merely isolation, as was seen in the belongingness narratives. The hero here is solely the cell phone as the primary source of something that humans apparently cannot provide for themselves or one another: stimulation. The key act in initiating this fantasy is the download.

The cell phone is becoming a virtual iPod (or Sony Walkman revisited), in the sense that while ostensibly a communication device designed for the purpose of connection, it is being increasingly marketed and defined as an entertainment center. This returns the user to an island of personal stimulation rather than interpersonal relationship or community. It also emphasizes corporate control, because the primary function becomes downloading—consuming—rather than relating. "Talk to me, I'm bored" could be a perfectly acceptable overture for conversation. It becomes increasingly difficult within ads at this level to distinguish product downloads from conversation "downloads." This progression is not simply an issue of technological convergence. Phones that do email and text messages, push-to-talk and cell calls, are examples of convergence that still emphasize the relational potential of cell phones to connect humans to one another. The growing emphasis on phones that stream and store music, movies, photographs of *Playboy* bunnies or *Cosmo* guys without shirts, games, TV shows and webisodes, and various pay-for-access services challenge that relational potential and emphasize the ongoing colonization of time and space by media corporations.

Conclusion

In Neil Postman's (1984) insightful classic *Amusing Ourselves to Death*, he notes that "entertainment is the supra-ideology of all discourse on television" (p. 87). As the cell phone becomes more of a multimedia device that includes visual data and "programming," the same ideology seems to be dominating the way the cell phone is marketed. By far the largest single category of ads that were collected in this sample were for downloadable music, videos, ring tones, and other such content, including a thirty-four-page Game Guide by Cingular. The implications of this shift are profound.

This chapter has offered an analysis of cell phone ads that demonstrates how basic and more elevated needs are active themes throughout both cell phone and cell phone plan advertisements. Each theme offers a fantasy theme or mini-narrative that implies scenes, characters, and ideal plotlines. Many of these themes are consistent with what might be called "everyday life," or at least an idealized version of it. Actors in the ads are hanging out at beaches, rushing to appointments or adventures, or doing a host of other activities with which many people can identify. However, by far the more dominant and powerful themes were those of transcendence that positioned the cell phone as a device that allowed users to move beyond traditional boundaries of time and space into a world of infinite potential within which they can be the center. Thus the cell phone becomes a device that potentially cuts users off from reality or community, rather than connecting them more deeply to it.

References

AAF SmartBrief. (2006, August 3). SmartQuote. Retrieved from http://www.smartbrief.com/alchemy/servlet/encodeServlet?issueid=605A1D07–6F76–4E66-AFA2-A5233797C685&lmid=null.

Borchers, T. A. (2005). *Persuasion in the media age* (2nd ed.). Boston, MA: McGraw-Hill.

Bormann. E. G. (1972). Fantasy and rhetorical vision: The rhetorical criticism of social reality. *Quarterly Journal of Speech*, 58, 396–407.

———. (1982). Fantasy theme and rhetorical vision: Ten years later. *Quarterly Journal of Speech*, 68, 288–305.

Brummett, B. (1991). *Rhetorical dimensions of popular culture*. Tuscaloosa: University of Alabama Press.

———. (1999). *Rhetoric of machine aesthetics*. Westport, CT: Praeger.

Burke, K. (1969). *A grammar of motives*. Berkeley: University of California Press.

Fiske, J. (1989). *Television culture: Popular pleasures and politics*. London: Routledge.

Frith, K. T. (1997). Undressing the ad: Reading culture in advertising. In K. T. Frith (Ed.), *Undressing the ad: Reading culture in advertising* (pp. 1–17). New York: Peter Lang.

Green N. (2003). Outwardly mobile: Young people and mobile technologies. In J. E. Katz (Ed.), *Machines that become us* (pp. 201–217). New Brunswick, NJ: Transaction.

Haddon, L. (2003). Domestication and mobile telephony. In J. E. Katz (Ed.), *Machines that become us* (pp. 43–55). New Brunswick, NJ: Transaction.

Hart, R. P., & Daughton, S. (2005). *Modern rhetorical criticism* (3rd ed.). Boston, MA: Allyn and Bacon.

Jagodzinski, J. (2004). *Youth fantasies: The perverse landscape of the media*. New York: Palgrave.

Katz, J. E. (2003). Do machines become us? In J. E. Katz (Ed.), *Machines that become us* (pp. 15–25). New Brunswick, NJ: Transaction.

Kraus E., & Auer, C. (2000). Introduction. In E. Kraus & C. Auer (Eds.), *Simulacrum America: The USA and the popular media* (pp. 1–20). New York: Camden House.

Kuypers, J. A. (2005). *The art of rhetorical criticism*. Boston, MA: Allyn and Bacon.

Messaris, P. (1997). *Visual persuasion: The role of images in advertising*. Thousand Oaks, CA: Sage.

Pierce, D. L. (2003). *Rhetorical criticism and theory in practice*. New York: McGraw-Hill.

Postman, N. (1984). *Amusing ourselves to death*. New York: Viking.

Rogers, E. M. (1995). *Diffusion of innovations*. New York: Free Press.

Rotzoll, K. B., & Haefner, J. E. (1996). *Advertising in contemporary society: Perspectives toward understanding* (3rd ed.). Chicago: University of Illinois Press.

Rushkoff, D. (1994). *Media virus: Hidden agendas in popular culture*. New York: Ballantine.

Shumann, M. (2006, July 17). Full year 2005 mega brands report. Retrieved August 22, 2006, from http://adage.com/abstract.php?article_id=110554.

Vitello, P. (2006, July 12). A ringtone meant to fall on deaf ears. Retrieved August 23, 2006, from http://www.nytimes.com/2006/06/12/technology/12ring.html?ex=1158724800&en=f737dec96ea38b21&ei=5070.

Weaver, R. M. (1970). *Language is sermonic*. Baton Rouge: Louisiana State University Press.

Yu, L., & Hou Tng, T. (2003). Culture and design for mobile phones for China. In J. E. Katz, (Ed.), *Machines that become us* (pp. 187–198). New Brunswick, NJ: Transaction.

· 3 ·

MOBILE TECHNOLOGIES AT WORK

· 10 ·

NETWORKS UNLEASHED

Mobile Communication and the Evolution of Networked Organizations

Calvert Jones and Patricia Wallace

Networked modes of organizations are flourishing in a broad spectrum of human activity, ranging from biotechnology research to transnational terrorism. Inexpensive information and communications technologies (ICTs) and their steady diffusion into everyday life have supported the growth of these networks among diverse, geographically dispersed populations. In the private sector, firms are relaxing hierarchical control in favor of looser, networked structures and open innovation (Chesbrough, 2003). Like-minded activists are coming together in more fluid networks to promote a common cause, with local, national, and transnational participation (Keck & Sikkink, 1998). More informally, kids are building and tweaking elaborate social networks through their adept use of mobile messaging services (Ling & Yttri, 2002). Although networks are not new phenomena, mobile ICTs have the potential to vastly increase their velocity, flexibility, and reach, albeit with limits and risks.

This chapter examines the role played by mobile communication technologies in the evolution of the networked organization. Drawing from research on the social implications of these technologies, it highlights ways in which their use may affect how networks function and develop. This chapter adopts a classic theoretical model for networked organizations (Powell, 1990), aiming to show how mobile communications influence and facilitate their core features.

Examples from practice illustrate the linkages between mobile communications and the networks in which they are embedded. The chapter also explores the risks and limitations of these linkages, revealing how they can undermine networked organizations as well as strengthen them.

The Networked Organization

In W. W. Powell's (1990) theory, the networked organization represents a mode of organization distinct from more familiar markets and hierarchies. Decentralized market exchanges tend to be impersonal, with information required to make decisions simplified and circulated through the price mechanism. A corn shortage due to poor weather or soil erosion, for example, is communicated to buyers through higher prices, without the potentially complex reasons behind the shortage made known to every buyer. Buyers can still make reasonable decisions on the basis of price information alone. Trust is thought to be less important because agreements can be enforced by law. As a mode of exchange, markets offer the benefits of flexibility and choice for the actors involved. In contrast, hierarchical organizations exert far more control over the exchange of information. Here, such exchanges are regulated by managers, routines, and procedures. Transactions are internalized within a clear structure of authority, and trust may be higher because of the guarantees provided by that authority. Information flow tends to be more cumbersome, having to travel up and down chains of command. Yet hierarchies offer their own advantages over the market, such as accountability and reliability.

The networked mode of exchange differs from markets and hierarchies in a number of ways. It brings together people collaborating for mutual benefit, interacting more informally than in a hierarchy and less impersonally than in a market. Networks emerge when one party depends on another party for resources, ranging from the more tangible, such as equipment, instructions, and material goods, to the more intangible, such as expertise, know-how, emotional reassurance, and ideological legitimacy. The exchange of these resources is voluntary and reciprocal, lacking both the discrete, isolated quality of market transactions and the more formal routines established by bureaucratic fiat. Communication is freer and lateral, with trust developing over time as exchanges prove their value to the parties involved.

The networked organization has several features thought to be advantageous in specific operating environments. It is more flexible than hierarchies, for example, because information flows more freely, unencumbered by narrow administrative channels and other restrictions, and participants are more autonomous. Networks can be more agile in responding to changing environments, with ideas circulating more rapidly and potentially translated into action more fluidly. Beyond their speed and flexibility, networks are thought especially suited to exchanging more intangible resources, such as expertise, intellectual know-how, technical skill, creativity, and innovative ability. These resources, with their more elusive, qualitative dimensions, are not well suited to either markets or hierarchies because they are hard to measure, codify, routinize, and price. They flow more easily through less formal networks of people engaged in recurring, complementary activities, in which trust and confidence build over time. Networks are well positioned to address complex, interdisciplinary problems because of their open-ended character, drawing in people of varied expertise, background, and ability.

The advantages of networked organizations have encouraged their emergence in fast-moving fields that deal with uncertainty and require diverse, frequently dispersed sets of resources. For example, activist groups seeking transnational participation are increasingly building networks with like-minded individuals and groups in other countries. After the December 2004 tsunami in Southeast Asia, when state-based disaster recovery efforts were seen as insufficient, volunteers worked in networks to provide ad hoc aid, drawing on people from around the world (Jones & Mitnick, 2006). In the rapidly developing biotechnology sector, interorganizational networks of learning are proving to be well suited for addressing research problems (Powell, Koput, & Smith-Doerr, 1996). Another example of an organization that leverages the advantages of the network is Al Qaeda. Reacting to the loss of their central headquarters in Afghanistan, these militants have evolved into a looser, less hierarchical global network (Gunaratna, 2002). Security experts are recommending that the U.S. intelligence community meet this threat by itself becoming more of a network, proposing that it takes a network to fight a network (Arquilla & Ronfeldt, 2001).

The evolution of networked organizations has relied heavily on connectivity, and the enabling power of ICTs. Email and a variety of collaborative technologies based on Internet access have made the growth of this form of organizational structure far easier than it would have been without them. Mobile technologies add another element, and the role that mobile communication plays in the evolution of these networks is discussed next.

Mobile Communication and the Networked Organization

Nothing about networks absolutely requires the use of advanced ICTs, let alone mobile communication. Yet these technologies have created the support platform for networks, eliminating much of the cost and friction involved in establishing this kind of organizational structure. Mobile ICTs have become key features of the global communications infrastructure, and to some extent they simply augment the power of existing telecommunications capabilities without adding new capabilities. A conference room, for example, might be equipped with Wi-Fi (a local area network that permits simultaneous wireless broadband access; see Jassem, chapter 2 in this volume) rather than Ethernet jacks, so attendees don't have to bring cables with them to reach their servers with their laptops. A cell phone will also serve many of the same purposes as a land line phone. Nevertheless, the mobile aspect of cell phones and laptops, combined with the many other technologies that take advantage of wireless connectivity, bring with them new capabilities and features that carry special meaning for the evolution of networked organizations.

Common sense indicates that such technologies will significantly overcome constraints of location and time, enhancing organizational mobility and reach. Yet the social science of technology adoption and diffusion tends to show a more nuanced, reciprocal relationship between users and technology, involving more give and take among disparate sets of users as opposed to straightforward cause and effect (Bijker, Hughes, & Pinch, 1987; Kline & Pinch, 1999). Technologies developed for particular purposes evolve as users reinterpret and adapt them to their own local contexts, sometimes in ways that are hard to foresee. While it is too early to say how the diverse array of mobile ICTs will affect networked organizations, a growing body of research on the social implications of these technologies points to several possibilities.

Facilitating Microcoordination

Social research on mobile phones describes their use for "microcoordination" of everyday life (Ling, 2004). Microcoordination refers to the real-time coordination of upcoming activities through direct contact among participants using mobile communications. Because it allows people to coordinate their activity while they are en route, it challenges coordination through reference to a fixed time. Instead of agreeing in advance to meet at a specific place and time, people can improvise their schedules more readily, adapting to exigencies as they arise. Researchers found that children, especially in Scandinavian countries where adoption of the technology is pervasive, were extremely skilled microcoordinators of social events (Kasesniemi & Rautiainen, 2002; Ling & Yttri, 2002).

The concept of microcoordination has several dimensions poised to build further flexibility into networked organizations (assuming participants are able to master the technologies involved as well as Scandinavian teenagers have mastered them). First, "midcourse adjustment" allows time to be used more efficiently. When a meeting is cancelled, participants can be told immediately of the cancellation, freeing them to redirect their energies elsewhere. Instead of traveling to the meeting place, waiting for other participants, ultimately finding out about the cancellation, and making the return trip, they can use this formerly "lost" time more productively. In addition, "iterative coordination" through mobile phones supports flexibility by letting participants progressively negotiate the circumstances of their upcoming activities. They can make general agreements about what they will do, "zooming in" on the specifics as the time approaches. Lastly, schedules are "softened" through microcoordination. Meeting agendas can be adjusted to accommodate participants who communicate in real time that they are coming earlier or later, by rearranging the order of topics and other aspects of organization.

Midcourse adjustment, iterative coordination, and softened scheduling may not rest easily in more hierarchical organizations, accustomed to stable routines and more central management. But these dimensions of microcoordination are well suited to more informal networked organizations, in which individual participants have more control over their collaborative activities and the exchange of resources. People in these types of organizations are better equipped to take advantage of this kind of finely graded coordination because of their greater autonomy and tendency to communicate more freely. They may also be more willing to identify and reap the benefits of iterative coordination and softened scheduling because their collaboration is voluntary and reciprocal. While iterative coordination might seem a hassle to unwilling or uninterested teammates, network participants may see it as an opportunity to adapt to shifting goals, environmental conditions, and their own changing needs.

The People Power II demonstrations in the Philippines, for example, leading to the overthrow of former president Estrada in January 2001, illustrate how networked modes of organization can take advantage of mobile phone microcoordination (Castells et al., 2004). Although the opposition to Estrada was not itself one cohesive "networked organization," it consisted of individuals and groups exhibiting networked collaboration toward a shared goal. Tens of thousands of demonstrators used mobile communication to coordinate and adapt their strategies of opposition between January 16 and January 20. Showing decentralized but effective leadership, participants sent and received text messages from friends about what to wear to make a collective statement and where to assemble next to have the most impact. Iterative coordination allowed them to take advantage of opportunities as they arose, and midcourse adjustment let them recover swiftly from counterdemonstration techniques. Their success, though resulting from many more factors than mobile communication, inspired a popular terminology to describe their tactics, including "smart mobs" and "swarming" (Rheingold, 2002).

Relaxing Boundaries and Extending Reach

The interactivity and reach of many mobile ICTs may also augment networked forms of organization by making boundaries more porous, enabling more broadly based participation and a more varied exchange of resources. A strong trend in research on communication technologies points to their ability to expand personal networks beyond the local level, supporting collective action on a larger scale through one-to-many broadcasting and other capabilities (Castells, 1996; Keck & Sikkink, 1998; Wellman, 2002). Keith Hampton (2003) argues that computer-mediated communication (CMC) facilitated collective action in an experimental suburban community outside of Toronto, called "Netville," by mobilizing weak ties. "Weak ties" (Granovetter, 1973) are loose connections between people whose relationship is casual and based on infrequent communication. Weak ties can build a more broad-based foundation for action by linking clusters of a community that would otherwise fragment.

Mobile ICTs are a step further in this direction, enhancing the potential for broad-scale collective action through their greater reach and widespread adoption. Networked organizations exhibit an open-ended, loose structure that can easily accommodate and engage new people, typically without bureaucratic constraints. Valuable weak ties may be more easily identified and tapped through their looser boundaries and volunteer ethos.

The volunteer networks that emerged to provide disaster relief following the 2004 tsunami in Southeast Asia, for example, illustrate how communication technologies can extend the boundaries for participation, especially when they are wireless and mobile. Within hours of the tsunami, volunteers around the world—many of whom had never met—worked through various forms of networked collaboration to assist in the disaster recovery effort (Jones & Mitnick, 2006). One of the most visible ad hoc efforts, the South-East Asian Earthquake and Tsunami Blog, attempted to identify resources, coordinate volunteers, and deliver these forms of assistance to afflicted regions. The volunteers used mobile ICTs not only to tap into a wider pool of assistance, but also to work directly with people in these regions whose wired connections had failed. Real-time communication with these regions, telling volunteers exactly what was needed and where, made it possible in principle to match specific victim needs with available resources. By rapidly drawing in new people with diverse skills and ways of assisting, this volunteer network gained access to an extraordinarily varied set of resources.

The disaster recovery example also shows the growing role mobile ICTs play for networks in circumstances in which land lines are unavailable. People in many developing nations, for example, rely heavily on cell phones because land lines are not reliable, making the technology far more than an added convenience. For example, microentrepreneurs in Rwanda are using mobile phones to extend their networks and make new business contacts. A study of their mobile phone use (Donner, 2005) found that roughly a fifth of their calls concerned such contacts and the new possibilities that opened up as a result of their expanded business networks. For networked organizations, the interactivity and reach of mobile communications can support a more varied exchange of resources by engaging new people, especially in areas with limited wired connectivity.

Tapping Just-in-Time Expertise

The extended reach offered by mobile technologies goes hand in hand with the ability to obtain expertise from a wider array of participants, whenever it is needed and wherever the expert happens to be located. Virtual teamwork, for example, which has become a staple in networked organizations around the world, involves members who are geographically separated. Being able to escape from geographical boundaries to create such teams has brought considerable freedom to organizations. While virtual teams experience a number of problems that collocated teams do not, they make it possible for an organization to share resources more effectively and efficiently (Handy, 1995; Jarvenpaa & Leidner, 1999; Wallace, 2004). An expert on copyright law in New York, for example, could participate in any number of teams with members anywhere on the planet.

Mobile technologies add a new dimension to the application of expertise in networks because it can be tapped "just-in-time." For a fast-moving project, the ability to contact an expert immediately may be critical and save time spent on unproductive work. The copyright attorney may be contacted during dinner, and the knowledge provided might prevent team members on the other side of the globe from wasting a day on a dead end.

Opening up Backchannels and Backstage Negotiation

Mobile communication also opens up more possibilities for "backstage" interaction, which can aid collaboration by providing a safe haven for questions, clarification, and tentative ideas to be aired. Erving Goffman (1959) uses a dramaturgical metaphor to differentiate between how people act in front of an audience, a kind of "front stage," and their more relaxed, informal behavior "backstage," away from the audience. In front regions, several people may work together to give a certain impression to an audience, such as the appearance of erudition, politeness, effortlessness, high class, diligence, and other desirable qualities, depending on the situation. They avoid actions and other signals inconsistent with this impression, as when waiters in top restaurants are coached not to display poor posture. Behind the scenes, they abandon the performance to relax, assemble "stage props," adjust their "costumes," resolve conflicts, test new approaches, and clarify or change procedures. The backstage region is commonly separated physically or by some other means from the front stage where the performance takes place, like a restaurant kitchen out of the view of diners.

Mobile ICTs provide further, more nuanced opportunities to engage in backstage interaction among networks of people. It is not necessary for the back region to be physically separated from the front region, as when students in a classroom chat through communication "backchannels" on their laptops, while maintaining their participation in the front region of the class itself. A study of graduate students chatting in an IRC backchannel during class found that they used it for a variety of purposes (Rothenberg & King, 2006). Many reported using it to clarify questions related to lecture content, including ones they were too embarrassed to ask aloud or which they felt were not relevant enough to warrant interrupting the class. They also used it to carry out related academic discussion, especially in situations when a student could supplement the lecture by giving more information and background on a topic. Some students valued the backchannel for exposing them to alternative perspectives in real time, as when a student might disagree with the professor and point to a URL with counterevidence. Apart from these more "productive" activities, students used the backchannel for entertainment and alleviation of boredom.

While backchannels might be seen as a subversion of authority or distraction in more hierarchical organizations, where the flow of information is more tightly controlled, they are a natural fit for informal, networked organizations with fewer rules and regulations. As opposed to the up-and-down path of information common in hierarchies, communication is freer in networks, making the concept of backchannels less threatening. The versatility of mobile ICTs, the ability to rapidly involve one or more people in a backchannel during a meeting or directly afterward, supports a wide variety of useful backstage interactions. Such backstage flexibility can contribute to collaboration by opening up "safe" venues for questions, clarification, and discussion of alternatives, involving any number of people. Backstage regions are important in work because they provide this type of refuge, where the front stage performance can be more informally negotiated and even improved. Indeed, workers face problems when they are unable to maintain sufficient control over back regions (Goffman, 1959). Mobile communication devices impart far more control over such backchannels, including which people are participating, when they are participating, and where.

Backstage interaction may also play a significant role in the process of innovation. Research on the innovative Silicon Valley region highlights how informal interactions, at bars and other back regions vis-à-vis the workplace, supported a rich flow of information, tips, expertise, and ideas (Saxenian, 1996). The researchers studying the graduate students' backchannel described one of its uses as a "testing ground" for ideas, which students might be too hesitant to broach in the "more official" front region of class (Rothenberg & King, 2006). In the backchannel, students could air tentative ideas, gauge a response, and refine them with input from peers before raising their hands in class. The safe haven element of backstage negotiation may therefore stimulate the exchange of a wider range of ideas and resources. Because of their informal, reciprocal character, networks are thought to be particularly conducive to the exchange of resources such as innovative skill, expertise, and know-how. Mobile communication may enhance this feature by opening up further opportunities for the informal backstage interactions where these resources flow more easily.

Risks and Limitations of Mobile Communication for Networked Organizations

Mobile ICTs, then, facilitate networked organizations in several ways. They add flexibility, and they can dramatically extend reach. They support and enable broader, more diverse participation and exchange of resources, and they provide venues for less formal communication and collaboration. The use of mobile technologies to support informal backchanneling is one such venue that may help spur innovation. While mobile ICTs offer many advantages, they also carry limitations and risks that complicate the relationship between mobile ICTs and networked collaboration.

Work, Nonwork, and Overload

With mobile ICTs making every member of the network available at any time, regardless of their physical location, the dividing line between work and nonwork has become increasingly blurred (Wallace, 2004). Expectations for "reachability" have grown, and coworkers expect to be able to contact one another during evenings, vacations, weekends, even in the middle of the night if necessary. While this capability is advantageous to networked organizations in many respects, supporting "just-in-time" expertise for project teams, for instance, it can add a heavy burden in the form of work stress. The flexibility that mobile ICTs bring to networked organizations bears a cost for work-life balance.

One reason mobile technologies heighten the risks for general overload is that they support the ability to fill in little time slots with work-related communication. Taking a taxi, standing in line, or waiting for an elevator were once opportunities to relax, if only for very brief periods. Cell phones and BlackBerry devices now make those time slots available for a bit of productive work. Those small time slots were not really long enough or convenient for reading or writing, but they are fine for brief text messages or phone calls. The result is that the short rest periods in which people could gaze about and relax, even while "on the clock," are less available.

Confusion and Credibility

Confusion, conflict, and duplication of effort may be other side effects of using mobile technologies for networked collaboration, such as when microcoordination involves larger groups. Although two people may easily engage in iterative coordination of an upcoming one-on-one meeting, the process grows far more complex when more people are involved (Ling, 2004). When one person makes an alternate suggestion, the message has to be sent to all other participants and their responses must in turn be received by all other participants. Microcoordination seems to work most efficiently in small groups, or small groups coordinating loosely with other small groups, when fewer details have to be negotiated.

Low credibility is a risk of mobile communication in the service of networked collaboration. Text messages, for example, are not yet widely accepted as a formal communication tool, partly because norms for their use are still emerging. It took many years for people to develop relatively stable norms about the use of email, and even now the medium can be a trying venue in which people easily blunder when they send mail, or misinterpret incoming messages (Wallace, 1999). Text messaging is an even more impoverished medium, and its credibility is not well established. Many associate it with teenager networks and the "thumb" culture. Broadcast text messages may be particularly problematic and perceived as spam or general advertising.

During People Power II in the Philippines, demonstrators were concerned that participants would not take text messages seriously, not necessarily knowing their origin and purpose (Castells et al., 2004). Religious organizations were therefore drawn in to build legitimacy into the text message coordination campaign. As one activist wrote in a listserv, "I was certain [texting] would not be taken seriously unless it was backed up by some kind of authority figure to give it . . . legitimacy" (Castells et al., 2004, p. 204). This activist told how Radio Veritas, a church-owned broadcasting station, helped add credibility to the waves of anti-Estrada text messages being sent out. The example reinforces more broadly the point that mobile ICTs do not exist in a vacuum; they interact with other social forces and communication media.

Security

Security is another significant risk associated with mobile ICTs that is heightened for networked organizations (see Williams, chapter 6 in this volume). The militant Al Qaeda network, for example, reportedly stopped using mobile phones for communication because they were so easily tracked (Gunaratna, 2002, p. 16). (In this case, the network did not abandon the technology, but reinterpreted and redefined its use. The mobile phone emerged as a means to both deceive intelligence services and detonate bombs remotely, rather than as a person-to-person communication device.) Additionally, the expanded reach of mobile ICTs, while facilitating broad-based participation and exchange of resources, can threaten security. In the case of an underground network like Al Qaeda, it may allow for broader recruitment efforts and interorganizational collaboration, but it can also open the door to informants and double agents (Jones, 2006). The openness of the South-East Asian Earthquake and Tsunami Blog raised security problems as well, as when scammers with fake charities called for help. Thus, while mobile ICTs can support wide-scale participation, they can also create problems of security and credibility.

Conclusions

Mobile ICTs offer both advantages and disadvantages to networked organizations and their participants. They support the looser, less hierarchical structure and informal communication characteristic of such organizations, and they bring new features that can be applied in innovative ways. Ultimately, however, the role that mobile technologies play will be shaped by how people actually interpret and use the technologies. Mobile backchannels, for example, can just as easily be a distraction from fruitful collaboration as they can be used to supplement and support it. They might be a useful venue for a parallel, private discussion of a conference presentation, or they might be used to play games and ward off boredom at the same presentation. Participants may choose to use them productively or otherwise.

Organizations, too, will make choices that will guide aspects of the way mobile technologies are used and applied, and how they affect organizational change. For example, will cell phone numbers be available so that all members of the organization have access to them? Email addresses became widely available over time, and the ease of sending messages to any (or every) member of an organization by email introduced capabilities that helped break down hierarchical boundaries and encouraged the development of the less formal networked modes of organization. Broadly distributed cell phone numbers may lead to further organizational change, as in expectations for availability and work-life balance.

Organizations will also grapple with the issue of location awareness. Cell phones do not, as a default, reveal the sender's location to the recipient. In fact, the location displayed, in the form of the area code on the caller ID, can be quite misleading because the devices are mobile. As we discussed, cell phones are used to microcoordinate physical gatherings, as people compensate for this location blindness and update one another on their current locations. The adoption of devices that routinely reveal their locations to the recipients of the communications, however, may have both positive and negative implications for privacy, security, productivity, and even morale.

The networked organization gained important capabilities when cell towers, satellites, microwave dishes, and other kinds of infrastructure emerged to support mobile technologies. The devices that take advantage of that infrastructure will continue to roll out and offer everything from simple voice communications and text messaging to global positioning systems (GPS) for navigation. Considering the historical role of technological innovation in social change, Karl Marx pointed out the close relationship between the development of the hand mill and feudalism, and between the rise of the steam mill and industrial capitalism. ICTs helped pave the way for loosely structured, networked organizations to emerge and thrive. The growing array of mobile technologies amplifies those capabilities, and adds new features that may drive further evolution and change. Yet social context will continue to condition this evolution, as networked organizations interpret and adapt innovations for their own divergent purposes.

References

Arquilla, J., & Ronfeldt, D. (Eds.). (2001). *Networks and netwars*. Santa Monica, CA: RAND.

Bijker, W. E., Hughes, T. P., & Pinch, T. (Eds.). (1987). *The social construction of technological systems: New directions in the sociology and history of technology*. Cambridge, MA: MIT Press.

Castells, M. (1996). *The rise of the network society, vol. 1*. Malden, MA: Blackwell.

Castells, M., Fernandez-Ardevol, M., Qiu, J. L., & Sey, A. (2004, October). The mobile communication society: A cross-cultural analysis of available evidence on the social uses of wireless communication technology. Paper presented at the Wireless Communication Policies and Prospects: A Global Perspective meeting, Los Angeles, CA. Retrieved August 15, 2006, from http://arnic.info/WirelessWorkshop/MCS.pdf.

Chesbrough, H. W. (2003). *Open innovation: The new imperative for creating and profiting from technology*. Cambridge, MA: Harvard Business School Publishing Corporation.

Donner, J. (2005). The use of mobile phones by microentrepreneurs in Kigali, Rwanda: Changes to social and business networks. Paper presented at the Wireless Communication and Development: A Global Perspective meeting hosted by the USC Annenberg Research Network on International Communication Workshop. Retrieved August 15, 2006, from http://arnic.info/workshop05/Donner%20_MobileKigali_Sep05.pdf.

Goffman, E. (1959). *The presentation of self in everyday life*. New York: Anchor.

Granovetter, M. (1973). The strength of weak ties. *American Journal of Sociology*, 78, 1360–1380.

Gunaratna, R. (2002). *Inside Al Qaeda: Global network of terror*. New York: Berkley Books.

Hampton, K. (2003). Grieving for a lost network: Collective action in a wired suburb. *Information Society*, 19, 417–428.

Handy, C. (1995). Trust and the virtual organization. *Harvard Business Review*, 73(3), 40–48.

Jarvenpaa, S. L., & Leidner, D. E. (1999). Communications and trust in global virtual teams. *Organization Science*, 10(6), 791–815.

Jones, C. (2006). Al Qaeda's innovative improvisers: Learning in a diffuse transnational network. *Cambridge Review of International Affairs*, 19(4), 555–569.

Jones, C., & Mitnick, S. (2006). Open source disaster recovery: Case studies of networked collaboration. *First Monday*, 11.

Kasesniemi, E., & Rautiainen, P. (2002). Mobile culture of children and teenagers in Finland. In J. E. Katz & M. Aakhus (Eds.), *Perpetual contact: Mobile communication, private talk, public performance* (pp. 170–192). Cambridge: Cambridge University Press.

Keck, M., & Sikkink, K. (1998). *Activists beyond borders: Advocacy networks in international politics*. Ithaca, NY: Cornell University Press.

Kline, R., & Pinch, T. J. (1999). The social construction of technology. In D. A. MacKenzie & J. Wajcman (Eds.), *The social shaping of technology* (2nd ed.) (pp. 113–115). Buckingham, England, and Philadelphia, PA: Open University Press.

Ling, R. (2004). *The mobile connection: The cell phone's impact on society*. San Francisco, CA: Morgan Kaufmann.

Ling, R., & Yttri, B. (2002). Hyper-coordination via mobile phones in Norway. In J. E. Katz & M. Aakhus (Eds.), *Perpetual contact: Mobile communication, private talk, public performance* (pp. 139–169). Cambridge: Cambridge University Press.

Powell, W. W. (1990). Neither market nor hierarchy: Network forms of organization. *Research in Organizational Behavior*, 12, 295–336.

Powell, W. W., Koput, K. W., & Smith-Doerr, L. (1996). Interorganizational collaboration and the locus of innovation: Networks of learning in biotechnology. *Administrative Science Quarterly*, 41, 116–145.

Rheingold, H. (2002). *Smart mobs: The next social revolution*. Cambridge, MA: Perseus Publishing.

Rothenberg, M., & King, J. (2006). Social uses of communication backchannels in a shared physical space. Unpublished manuscript, University of California at Berkeley. Retrieved October 11, 2006, from http://groups.sims.berkeley.edu/backchannel/downloads/backchannel.pdf.

Saxenian, A. (1996). *Regional advantage: Culture and competition in Silicon Valley and Route 128*. Cambridge, MA: Harvard University Press.

Wallace, P. (1999). *The psychology of the Internet*. Cambridge: Cambridge University Press.

———. (2004). *The Internet in the workplace: How new technology is transforming work*. Cambridge: Cambridge University Press.

Wellman, B. (2002). Little boxes, glocalization, and networked individualism. In M. Tanabe, P. van den Besselaar, & T. Ishida (Eds.), *Digital cities II: Computational and sociological approaches* (pp. 10–25). Berlin: Springer-Verlag.

· 11 ·

MEDICAL COMMUNICATION

Improving Patient Safety in the Operating Room and Critical Care Unit

Keith J. Ruskin

Modern communication technology is changing the practice of medicine. Physicians, nurses, and other healthcare professionals are by nature mobile. They do not have fixed workspaces, but instead move from patient to patient, making them difficult to contact quickly. Each member of the healthcare team must communicate with colleagues even when they cannot all be at the patient's bedside. Until recently, physicians could only be reached by pager, and they could not return a call unless they were close to a hospital extension or an outside telephone. Receiving a page while driving would result in a desperate search for the nearest pay telephone. Mobile telephones, two-way pagers, and instant messaging have made it easier than ever for co-workers to contact each other and for patients to contact their doctors.

Patients in the operating room or critical care unit require the constant attention of a diverse group of healthcare professionals, including doctors, nurses, and therapists. Information such as laboratory data, imaging studies, and procedure schedules is generated throughout the healthcare process, and must be quickly and accurately transmitted to physicians and allied health professionals. Efficient and reliable communication in the clinical environment is therefore critical to patient safety. Modern communication tools such as cellular telephones and wireless computers can dramatically improve patient care by providing rapid, efficient access to critical information.

Healthcare Information

Physicians and nurses rely on many kinds of information while caring for their patients. Verbal conversations may be ideal for some applications, such as requests for a specialty consultation or sharing a plan for diagnosis or treatment. Healthcare providers may, however, also need access to numerical data, pictures, video, or physiologic waveforms in order to make a decision. The advent of both wired and wireless broadband connections, combined with ubiquitous cellular telephone service, has enabled physicians to participate in patient care decisions regardless of their physical location (Cermak, 2006; Kyriacou et al., 2003).

Numerical information, such as laboratory values and vital signs (heart rate, blood pressure, and respiratory rate), is the basis of many decisions. For example, a normal white blood cell count might mean that a patient can be discharged from the hospital. This patient could be released from the hospital more quickly if the information could be sent to the patient's physician and nurse and to administrative personnel responsible for assigning beds. Some medications are adjusted to maintain the blood level within a narrow therapeutic window. Clinical alerts of abnormal drug levels enable physicians to modify the dose of the drug as soon as the laboratory value becomes available. Without such a system, several hours might elapse before the physician could access a computer or contact the laboratory staff. Clinical alerting systems that automatically send alphanumeric pages with abnormal values have been shown to improve patient care and are being installed in a growing number of hospitals.

Vital signs and waveforms are usually monitored by nurses either at the patient's bedside or at a central station in an intensive care unit. Blood pressure as measured from heartbeat to heartbeat, heart rate, and blood oxygen saturation are essential to the management of critically ill patients. Parameters such as waveforms from catheters placed in the heart and respiratory gas analysis may also help to guide patient care. Although telemetry, or the remote monitoring of patient vital signs, has been available for many years, the advent of wireless networks and portable handheld computers can significantly improve access to this critical information.

Modern patient monitors are capable of transmitting both numbers, such as heart rate and waveforms, such as an electrocardiogram or blood pressure tracing. Although skilled personnel should remain close to a critically ill patient's bedside, the physician who is ultimately responsible for that patient can perform other duties while being immediately available in the event of a sudden change in the patient's condition. The ability to transmit a live video image of the patient makes it possible for a remote physician to guide others at the bedside through evaluation and treatment of many problems. This technology has allowed the development of remote ICU monitoring that provides patients at remote hospitals with the expertise of highly trained critical care physicians (Breslow et al., 2004).

Physicians have also begun to help their colleagues perform complex procedures using remote telementoring. Telementoring involves the transmission of video images, pictorial information (such as X-rays or magnetic resonance imaging [MRI] scans), and physiologic data to experienced personnel at a tertiary care center, who then give instructions to the physician who is actually performing the procedure. Radiologists have been among the first physicians to use telementoring, and there have been several reports that prove its effectiveness. It has also become common practice for radiologists to interpret X-rays and scans from home using specialized software and a high-speed Internet connection, providing critical diagnostic information as soon as the images are available. This technology allows highly specialized physicians to provide advanced medical care for patients who may not be able to travel to tertiary care centers (Di Valentino et al., 2005).

Communication Technology

Modern communication tools allow multiple forms of information to be sent by voice, by short alphanumeric messages, through real-time access to computer networks, and by video. Each of the devices that can be used in the clinical setting has specific advantages and disadvantages.

Pagers were introduced over three decades ago and continue to be widely used in the healthcare setting. Paging systems are inexpensive, simple to install and use, and offer a high level of reliability. The lag time between transmission and receipt of the page may be only a few seconds, especially when dedicated, in-house systems are used. Although the death of paging has been predicted many times, most technology experts recommend that hospitals retain paging systems for extremely urgent messages. Radio paging remains the most efficient mechanism for immediate distribution of alerts to multiple people, since most other technologies send a message to one person at a time (Heslop et al., 2003).

Alphanumeric pagers are used by physicians in office practice, and in nearly every hospital and ambulatory care facility. Although most pagers can display only a short message, they improve the efficiency of patient care by allowing laboratory results or other clinical alerts to be transmitted to physicians. Automatic paging can decrease the lag between the time the test is completed and when the results can be transmitted to the clinician (Kuperman et al., 1996). Such a laboratory alerting system has been implemented in a large teaching hospital (Poon et al., 2002). The clinician can request automated paging for both normal and abnormal laboratory tests for a specific patient. Alphanumeric pagers are also an integral part of at least one operating room management system. The GE Centricity system is capable of automatically alerting the anesthesiology and nursing teams when certain surgical milestones are reached or when the surgical schedule changes.

Despite their widespread use throughout the healthcare industry, pagers have significant disadvantages that limit their utility in the patient care setting. Most pagers permit only one-way transmission of a telephone number or short text message. There is no way to respond through the pager or even acknowledge receipt of an important message. As a result, the person who receives the page must usually find a telephone and return the call. If the number is busy, or an incorrect telephone number was entered into the paging system (as may happen when the person paging is distracted during an emergency), a further delay in communication may result.

Mobile telephones provide rapid, two-way communication by voice as well as with pictures and short text messages. As a result, users can exchange information much more efficiently than they can with pagers. Mobile telephones also allow an important message to be discussed or at least acknowledged. They also allow important information to be shared through the use of text and picture messaging. By providing high-speed Internet access, mobile telephones can be used to access hospital information systems wherever there is service. For these reasons, mobile telephones have been rapidly adopted by nearly every other major industry, including emergency services, transportation, manufacturing, and the public sector, as a primary mode of communication.

Modern mobile telephones use "cellular" technology to provide high-quality communications using very low levels of emitted radiofrequency energy. This technology divides metropolitan areas into small loci, or cells, each of which is equipped with a low-power transmitter and sensitive receiver on a tall tower. Digital cellular telephones make use of several different standards that increase efficiency and allow many handsets to share a single frequency. The most commonly used protocol in the United States is called code-division multiple access (CDMA), which allows the base station to control the transmitter power used by the handset. In most other parts of the world, mobile handsets use the global standard for mobile communications (GSM). Cellular telephones offer several advantages over other communication technologies. They are inexpensive and widely available, and handsets using the GSM standard can work nearly anywhere in the world. Cellular telephones may not work well inside large buildings, however, and pose a very small risk of electromagnetic interference.

Wireless local area networks (WLANs) were introduced several years ago as a means of creating local area networks without the need for stringing wires from computer to computer. Although originally intended as a means for computers to share data and peripherals, the technology is becoming increasingly popular as a way to facilitate voice and data communication. Wireless networks typically have the capacity to handle multiple data streams at once, are relatively inexpensive to set up and maintain, and are highly versatile. As an added benefit, a device on a wireless network can be handed off from access point to access point, providing seamless coverage over a large area. The amount of radiofrequency energy generated by WLANs is very low, and the frequencies used are well removed from those used by medical devices. These advantages, combined with the low cost and wide availability of WLAN equipment, make this technology well suited for use in many healthcare institutions.

Computer networks can carry voice conversations by digitizing the audio signal and transmitting it across the network to any device with compatible software. This technology is called voice over Internet protocol (VoIP) and can be used to transfer a call through the public telephone. It is widely accepted by the general public as an alternative to traditional telephone service, and specialized versions are now being installed in large hospitals. A new communication system developed by Calypso Wireless uses portable telephone handsets that seamlessly switch from VoIP when WLAN service is available to cellular telephone service when carried outside of the building. This offers the potential to allow a single telephone to work as an "in-house" hospital extension when it is inside the building. The system automatically routes calls over the public network to that same telephone when it is removed from the building.

Some healthcare institutions are evaluating VoIP systems as an inexpensive, highly effective communication technology. Vocera Communications provides wearable, voice-controlled communication badges that can be clipped to a pocket or worn around the neck. Badges enable users to communicate with each other over a WLAN network. Users can also place calls over the public telephone network through a dedicated server that is integrated with a telephone exchange. The Vocera system allows users to contact each other by name (for example, "Call Keith Ruskin"), role ("Call attending anesthesiologist, room 9"), or group ("Call Cardiac Anesthesia"). This system has been used in several hospitals, but evaluations were mixed. In one study, the speed of communication to and from anesthesia providers was found to be four times faster than that provided by conventional pagers. Seventy percent of the personnel using that system preferred it to alphanumeric pagers and telephones. In another institution, the use of VoIP was limited by "dead spots" and questions about confidentiality of health information (Jacques et al., 2006).

The development of widely used standards has made it possible to connect medical equipment together and to transmit the data that they generate through existing data networks to desktop computers, servers, or monitoring stations. For example, healthcare professionals routinely need access to cardiac telemetry, laboratory data, hospital information systems, paging, and telephone networks. Until recently, each system used its own network of transmitters, receivers, and cabling and required a dedicated device to view the information. Most new equipment uses international standards for transmitting, storing, and displaying information, meaning that information can frequently be retrieved from personal computers, laptop computers, and personal digital assistants (PDAs) with wireless network access. The end result is that a core communication infrastructure can be installed using readily available equipment, and that this infrastructure is likely to be compatible with new devices.

Interference with Medical Equipment

Hospitals in the United States and Europe have implemented policies that prohibit the use of wireless communication devices in patient care areas (Klein & Djaiani, 2003). Unfortunately, most of these policies were developed in response to anecdotal reports of interference and ignore the potential benefits that cellular telephones can bring to patient care. These policies were written in response to published case reports and early studies in which mobile telephones were suspected of causing malfunctions in physiologic monitors and life-support devices. Anecdotal reports of interference were submitted to the Centers for Devices and Radiological Health and then forwarded to the U.S. Food and Drug Administration.

In one large study of mobile telephones in the operating room, the reported prevalence of interference is 2.4 percent, which is much less than the risk of medical error or injury resulting from a delay in communication, 14.9 percent (Soto et al., 2006). This small risk must, however, be weighed against the potential risk of interference with life-support devices such as mechanical ventilators, infusion pumps, and monitoring equipment. There were no life-threatening events reported in this study, and data from other studies support this finding. ECRI (formerly known as the Emergency Care Research Institute), a non-profit health services research agency and a Collaborating Center of the World Health Organization, has recognized the improvement in cellular technology and currently recommends the use of mobile telephones when required for rapid clinical communication. Most experts recommend that mobile telephones and wireless computers should, however, be kept more than three feet from medical devices (Shaw et al., 2004).

The infrequent reports of electromagnetic interference from cellular telephones are probably attributable, in part, to improvements in cellular telephone technology. Three factors determine whether the energy that is emanated by wireless communications devices will cause disturbances in medical equipment: power, shielding, and proximity. The power output of a typical handheld cellular telephone is very low, typically much lower than that used by a handheld radio or walkie-talkie. Institutions that make use of cellular telephones in clinical areas can install "microcells" that provide dense signal coverage while causing mobile telephones in the area to use the minimum power output. Some areas within hospitals, especially operating rooms, are electrically noisy or are located away from outside walls. Installation of microcells also permits mobile telephone use in these areas in which coverage would otherwise be poor.

Cellular telephones and wireless networks use frequencies that are well removed from those used for medical telemetry equipment. Modern telemetry equipment uses an assigned frequency range that is not near those used for wireless communication devices. Modern handsets (and wireless networks) also use very low power, and most new medical devices are shielded to prevent entry of unwanted radiofrequency energy. As technology improves, the risk of interference will decrease. Unfortunately, policies regarding communication technology in the healthcare environment frequently do not keep pace with other industries.

The Massachusetts General Hospital has adopted a practical and evidence-based approach to the issue of personal communication devices. This policy dictates that devices such as cellular telephones and wireless handhelds may be used safely in most areas of the hospital and near life-support and diagnostic laboratory equipment subject to certain limitations. Untrained staff and employees, patients, visitors, and outside personnel are not allowed to use wireless communication devices closer than three feet, or roughly one arm's length, from any medical device or diagnostic laboratory medical device. Communication devices may be left in "standby mode" (that is, they do not need to be turned completely off) when located within three feet of a medical device, but the user must move the device outside the three-foot limit to use it. For example, if a cellular telephone rings, the user should move the cell phone away from the medical devices before answering the call. Because they use higher power, handheld radios (that is, walkie-talkies) must not be used within ten feet of any medical device (Massachusetts General Hospital, 2004). These guidelines allow healthcare personnel to reap the benefits of modern information and communication technologies while minimizing the risk of interference with life-support equipment.

Communication Standards

It is important that medical devices and individual medical record systems as well as secure third-party storage and retrieval systems be accessible to any healthcare provider with a legitimate need for the information. Open standards promote cooperation among different sectors of the healthcare industry and should be used throughout the system. The rapid growth of the Internet and the development of new networking technologies not envisioned by the creators of the Internet Protocol demonstrate the value of nonproprietary standards. Several groups currently develop and maintain standards for the interoperability of healthcare information systems.

The Systematized Nomenclature of Medicine, known as SNOMED, is a standardized language for electronic health records maintained by the College of American Pathologists. SNOMED consists of a list of terms and their definitions, and allows any document to be indexed or searched according to specific criteria. Although it started as a common language for pathology, SNOMED now serves as a data dictionary for multiple medical specialties and dentistry. Other specialties are allowed to develop their own terminology, which can then be incorporated into the SNOMED standard. For example, the Data Dictionary Task Force of the Anesthesia Patient Safety Foundation has been developing a common language for anesthesiology. The Health Level Seven (HL7) organization has been accredited by the American National Standards Institute to develop standards for exchange and management of data that support clinical patient care. Its language, also called Health Level 7 (HL7), is designed to facilitate administrative functions such as admissions, discharges, and patient transfers.

Extensible Markup Language (XML) is used to create documents that contain structured information. It is widely used throughout many industries to describe and identify specific kinds of data, and makes it possible for two companies using independently developed programs running on incompatible systems to share information such as inventory or order status. These characteristics make XML a flexible and efficient way to define the structure of the information within the network. Within the healthcare industry, the HL7 clinical document architecture project makes use of XML to define the characteristics of each item in the medical record. Using XML to exchange information can potentially allow a computer program to automatically determine the structure of a newly added database and then search for information that is required.

Privacy and Security

Patients expect that their medical records will remain confidential. Government regulations as well as ethical obligations require that patients' health information be protected whenever it is aggregated, stored, or transmitted. Most of the requirements for storage and transmission of medical records are covered under the Health Insurance Portability and Accountability Act of 1996 (HIPAA). These requirements extend beyond health professionals who collect information. Any provider of services to a healthcare organization that handles "protected health information" is bound by HIPAA to defend the security of medical records.

HIPAA regulations mandate both physical and electronic protection of health records. The requirements for electronic health records are more stringent than those for paper records, and are applied to information that resides in a single computer or on a server, or that is transmitted across a network. The regulations also apply to letters, laboratory results, and even telephone conversations. Health information must therefore be encrypted before it is transmitted over public networks. It is also the responsibility of healthcare workers who are discussing a patient over the telephone to verify the identity of all participants in the conversation.

One potential solution that may be used to ensure privacy and security of digital information is to encrypt the medical record. A "smart card" carried by the patient would contain a key that permits access to the medical record. Providing the smart card would allow access to specific areas of the record for a period of time defined by the type of encounter. Under such a system, a clinical laboratory worker might be granted one-time, write-only access to the record unless a specific test required additional privileges. If the patient is admitted to the hospital, read and write access would be granted to authorized personnel for the duration of his or her stay. A primary care provider could be granted unlimited read and write access for the duration of his or her relationship with the patient. Records of encounters with mental health professionals would be in a separate, highly confidential category and require specific permission to access. The patient could also choose to restrict access to specific portions of the medical record. In the event of a medical emergency, information would be made available to the clinician through the use of a third key that would be administered at the regional or national level. Taiwan is one of the first countries to implement the use of smart cards as part of a national health system. Although implementation of the system was complicated by problems with the card readers and lack of familiarity with the system, the majority of hospitals in Taiwan were satisfied with smart cards as a way to gain access to patient records (Liu et al., 2006).

Attacks on personal computers in the form of viruses, keystroke loggers, and "phishing" attacks are a growing threat and have the potential to interfere with patient care. (See Williams, chapter 6 in this volume, for a discussion of cyber-crime.) Hundreds of viruses are released every day. Many healthcare applications rely on Intel-based computers running the Microsoft Windows operating system. As a result, they are vulnerable to the same kinds of viruses that affect home and office computers. Some experts have suggested that terrorists may specifically target the information infrastructure in hospitals and clinics to increase the number of casualties during an attack. In addition to rendering a computer unreliable, viruses and worms can compromise or destroy health information. Information networks are part of the critical infrastructure of most healthcare institutions and should therefore be protected. Fortunately, electronic and physical protection of critical infrastructure is a mature industry, and most healthcare institutions have taken steps necessary to secure their data.

Patient Safety

The Institute of Medicine estimates that medical errors are the eighth leading cause of death in the United States, and may cause as many as 100,000 patient deaths per year. To date, most medical research has focused on understanding disease processes or developing novel treatments. However, a significant number of medical errors have been attributed to a single cause: failure to communicate accurately. Some patient care areas, such as the operating room and intensive care unit, can be noisy and confusing, which may contribute to mistakes that occur as a result of a lapse in communication (Liu & Tan, 2000).

Healthcare workers spend much of their time communicating with each other, and a lag between the time a decision is made and action can be taken further increases the risk of error. Not only does communication consume a significant amount of time, but failure to communicate has been shown to be a root cause of medical errors, and in one study was shown to be the second most prevalent cause of medical errors (Hersch et al., 2002; Kluger & Bullock, 2002). Another study revealed that operating room nurses make up to seventy-four communication efforts per hour (Hersh et al., 2002). An error due to a lapse in memory may occur when as little as ten seconds separate the communication from the required response (Parker & Coiera, 2000). Instantaneous, accurate communication may, therefore, improve patient safety.

Effective communication has been shown to be a critical component of safety in high-risk environments and is essential to patient safety. Several reviews have postulated that improving communication among healthcare professionals may improve patient safety. The results of at least one study suggest that the use of mobile telephones decreases the incidence of errors (Soto et al., 2006). Facilitating communication by installing a comprehensive infrastructure that permits verbal communication as well as the exchange of pictures and physiologic data may, therefore, reduce the risk of medical mistakes.

Health Information Technology and Communication

Information technology has become an integral component of healthcare. The rapid growth in the variety and quality of online medical information offers all physicians an unprecedented opportunity to use information technology for both education and patient care. Clinical information, including continuing medical education and the latest journal articles, can be retrieved from anywhere in the world. Laptop computers and handheld devices have become smaller, faster, more rugged, and easier to use. Combining these new devices with innovative applications has already revolutionized patient care.

Many hospitals and ambulatory surgery centers now store patient information in electronic health records (EHRs), which provide access to patient information throughout the hospital and in physicians' offices. Journal articles and clinical guidelines can be incorporated into the EHR as part of a decision support system that can help healthcare providers to choose the best course of therapy for each patient. Widespread implementation and use of electronic clinical information systems, particularly electronic health records, provide a unique opportunity to improve the quality and efficiency of healthcare delivery.

A proposed National Health Information Network (NHIN) has the potential to allow any authorized healthcare professional involved in the care of a patient to access any medical information that had previously been generated for or on behalf of that patient. The Massachusetts E-Health Collaborative is one potential model of how health information will be shared in the future. The goal of the Collaborative is to make medical information widely available to healthcare professionals in Massachusetts. The Collaborative preserves confidentiality through the use of secure information systems. The initiative is not limited to transmitting the medical record from place to place; it is designed to facilitate the use of decision support systems and may in the future be expanded to include reimbursement information.

Conclusions

Clinical care often requires rapid communication among mobile healthcare providers. As doctors and nurses are asked to provide more sophisticated healthcare in hospitals as well as remote locations throughout the community, they will become increasingly reliant on efficient, reliable telecommunication. Because beeper systems can usually only send one-way messages, they may not be the best way to make contact with members of the care team during an emergency. They are, however, a good way to send short messages like a warning about an abnormal laboratory value. Recent alternative technologies, such as cellular telephones and handheld computers with wireless, provide a potentially advantageous shift from one-way communication to synchronous, two-way communication.

Modern communication technology makes it possible to help colleagues in remote locations to provide safe, efficient care to patients in the operating room and intensive care unit. The dynamic nature of the healthcare environment makes it an ideal place to demonstrate the benefits of modern communication tools.

References

Breslow, M. J., Rosenfeld, B. A., Doerfler, M., Burke, G., Yates, G., Stone, D., et al. (2004). Effect of a multiple-site intensive care unit telemedicine program on clinical and economic outcomes: An alternative paradigm for intensivist staffing. *Critical Care Medicine*, 32(1), 31–38.

Cermak, M. (2006). Monitoring and telemedicine support in remote environments and in human space flight. *British Journal of Anaesthesia*, 97(1), 107–114.

Di Valentino, M., Alerci, M., Bogen, M., Tutta, P., Sartori, F., Marty, B., et al. (2005). Telementoring during endovascular treatment of abdominal aortic aneurysms: A prospective study. *Journal of Endovascular Therapy: An Official Journal of the International Society of Endovascular Specialists*, 12(2), 200–205.

Hersh, W., Helfand, M., Wallace, J., Kraemer, D., Patterson, P., Shapiro, S., et al. (2002). A systematic review of the efficacy of telemedicine for making diagnostic and management decisions. *Journal of Telemedicine and Telecare*, 8, 197–209.

Heslop, L., Howard, A., Fernando, J., Rothfield, A., & Wallace, L. (2003). Wireless communications in acute health-care. *Journal of Telemedicine and Telecare*, 9, 187–193.

Jacques, P., France, D., Pilla, M., Lai, E., & Higgins, M. (2006). Evaluation of a hands-free wireless communication device in the perioperative environment. *Telemedicine and e-Health*, 12(1), 42–49.

Klein, A., & Djaiani, G. (2003). Mobile phones in the hospital: Past, present, and future. *Anaesthesia*, 58, 353–357.

Kluger, M. T., & Bullock, M. F. (2002). Recovery room incidents: A review of 419 reports from the Anaesthetic Incident Monitoring Study (AIMS). *Anaesthesia*, 57, 1060–1066.

Kuperman, G. J., Teich, J. M., Bates, D. W., Hiltz, F., Hurley J., Lee, R., & Paterno, M. (1996). Detecting alerts, notifying the physician, and offering action items: A comprehensive alerting system. Proceedings: A conference of the American Medical Informatics Association, AMIA Annual Fall Symposium, 704–708.

Kyriacou, E., Pavlopoulos, S., Berler, A., Neophytou, M., Bourka, A., Georgoulas, A., et al. (2003). Multi-purpose healthcare telemedicine systems with mobile communication link support. *Biomedical Engineering Online*, 24(2), 7.

Liu, E. H., & Tan S. (2000). Patients' perception of sound levels in the surgical suite. *Journal of Clinical Anesthesia*, 12, 298–302.

Liu, C. T., Yang, P. T., Yeh, Y. T., & Wang, B. (2006). The impacts of smart cards on hospital information systems: An investigation of the first phase of the national health insurance smart card project in Taiwan. *International Journal of Medical Informatics*, 75(2), 173–181.

Massachusetts General Hospital. (2004, December). Massachusetts General Hospital internal policy on personal communications devices.

Parker, J., & Coiera, E. (2000). Improving clinical communication: A view from psychology. *Journal of American Medical Informatics Association*, 7, 453–461.

Poon, E. G., Kuperman, G. J., Fiskio, J., & Bates, D. W. (2002). Real-time notification of laboratory data requested by users through alphanumeric pagers. *Journal of the American Medical Informatics Association*, 9(3), 217–222.

Shaw, C. I., Kacmarek, R. M., Hampton, R. L., Riggi, V., El Masry, A., Cooper, J., & Hurford, W. E. (2004). Cellular phone interference with the operation of mechanical ventilators. *Critical Care Medicine*, 32(4), 928–931.

Soto, R., Chu, L., Goldman, J., Rampil, I., & Ruskin, K. (2006). Communication in critical care environments: Mobile telephones improve patient care. *Anesthesia Analgesia*, 102(2), 535–541.

· 12 ·

THERAPY AT A DISTANCE

Information and Communication Technologies and Mental Health

Penny A. Leisring

Information and communication technologies (ICTs), such as computers and cell phones, are altering how and where many therapists and clients interact. With a few keystrokes and clicks of a mouse, a would-be client searching the Internet for "online therapy" can easily locate clinicians providing Web-based services. The implications are far-reaching and dramatic: clients who might not otherwise seek face-to-face treatment may be more willing to enter treatment in an online format. Also, clients for whom traditional face-to-face treatment is not an option (for example, some physically disabled individuals, people with agoraphobia, and families or couples separated geographically) can now receive services. Treatments via email, discussion boards, chat rooms, and videoconferencing are considered effective alternatives for underserved, confined, or isolated people (Jerome et al., 2000). These treatments are called telehealth, behavioral telehealth, cybertherapy, computer-mediated therapy, Web-based therapy, e-therapy, or Internet-based therapy.

While telehealth interventions are not currently a significant part of clinical practice for mental health problems (Murphy, 2003), they are slowly becoming more common, and a research literature examining their efficacy is burgeoning. (For an interesting history of e-therapy, see Skinner & Zack, 2004.) Therapy at a distance is not new; telephone counseling programs have been in existence since the 1950s (Hornblow, 1986), and anonymous crisis telephone hotlines are abundant for problems such as suicidal thoughts, rape, domestic violence, smoking, gambling, and more. Maxine Rosenfield and Evelyn Smillie (1998) conducted a small therapeutic telephone support group for women with various cancers. Shu-Hong Zhu, Gary Tedeschi, Christopher Anderson, and John Pierce (1996) describe an individual phone counseling program for smoking cessation. More recently, Alina Morawska and Matthew Sanders (2006) described a predominantly self-help behavioral parent training program that incorporated minimal therapist telephone contact.

Computers, considered useful for self-assessment and self-monitoring for some time (Kenardy & Adams, 1993), are now being used along with the Internet for psychological interventions—but not without much controversy. While using ICTs to augment traditional face-to-face therapy is somewhat accepted, many people, including clinicians, are wary of interventions administered purely online. Initially, Web-based treatment programs were evaluated only in terms of clients' and clinicians' satisfaction. However, controlled treatment outcome studies are now being conducted with promising results for many presenting problems. This chapter discusses research supporting Web-based mental health interventions and describes the many advantages and disadvantages of conducting these interventions. Legal and ethical issues are addressed. Psychotherapy supervision at a distance, called telesupervision, is also discussed.

What Does Therapy Look Like When It Is Conducted with ICTs?

Many types of computer-mediated communication can be used for therapeutic purposes. Perhaps the most common is email interaction between a therapist and client. With a quick Internet search, potential clients can find Web sites where they can choose to ask a therapist one question, a series of questions, or enter into an ongoing email relationship with a therapist.

Email can also be used as an adjunctive service along with ongoing traditional face-to-face therapy. Janice Murdoch and Patricia Connor-Greene (2000) describe two cases in which email was used to facilitate therapeutic homework assignment completion between face-to-face sessions for college students. In these cases, email was used for communication between sessions and for extra practice and feedback. Murdoch and Connor-Greene believe that email can enhance the therapeutic relationship and may enhance the effectiveness of treatment because clients regularly think about their behavior and treatment goals between sessions and get extra feedback.

Joel Yager (2001) echoes this sentiment and describes how email was used to supplement weekly face-to-face treatment for people with anorexia nervosa. By increasing the frequency and amount of communication, he feels that clients get the sense that the therapist is listening and thinking about them. Yager postulates that reporting via email between sessions about behaviors such as caloric intake may free up time during face-to-face sessions to discuss other issues. It has also been suggested that facilitating email communication between similarly diagnosed clients within a clinical practice may be beneficial and increase support for clients (Sansone, 2001).

E-therapy methods using discussion boards and scheduled chat groups are also being used (Glueckauf & Ketterson, 2004). People can receive support from peers or clinicians synchronously or asynchronously (Griffiths, 2005). Discussion boards allow people to post and read messages whenever it is convenient for them, while chat groups allow for more spontaneous discussion among members participating in the chat at the same time. Web sites can also be used to disseminate information (Castelnuovo et al., 2003).

Kristine Luce, Andrew Winzelberg, Megan Osborne, and Marion Zabinski (2003) theorize that Internet programs could be used for relapse prevention to prompt clients for information and to provide feedback to clients. They use the example of eating and weight issues: clients could be prompted to self-report information about caloric intake and exercise and then could be provided with tailored feedback and informed when they should contact a mental health professional.

The list of populations being treated using e-therapy is growing exponentially, including: isolated populations (Capner, 2000), physically disabled people (Hopps, Pépin, & Boisvert, 2003), military personnel overseas (Jerome et al., 2000), prisoners (Nickelson, 1998), people with behavioral medicine issues (Glueckauf & Ketterson, 2004), as well as those with anxiety disorders (Newman, 2000), encopresis (Ritterband et al., 2003), eating and weight issues (Abascal et al., 2004; Tate, Wing, & Winett, 2001), problem drinking (Postel, de Jong, & de Haan, 2005), and schizophrenia (Rotundi et al., 2005). Caregivers of people with schizophrenia (Rotundi et al., 2005), Alzheimer's disease (Glueckauf et al., 2005), and traumatic brain injury in childhood (Wade et al., 2005) have also been targeted.

Marloes Postel, Cor de Jong, and Hein de Haan (2005) indicate that Web-based therapy seems to attract a different population than traditional services. They found that those who entered an e-therapy program for problem drinking were more likely to be female, employed, and older than those seeking traditional treatment. Robert Glueckauf and colleagues (2005) suggest that the fit between a population and the technology used needs to be examined. They administered services for people caring for patients with Alzheimer's disease using the telephone instead of the Internet, because much of their target population is elderly and might not have the necessary computer skills or equipment.

Evidence Supporting Therapy at a Distance

While initial research of online therapy basically assessed clinician and client satisfaction with services, results from controlled research studies comparing groups receiving online services to control groups receiving no services are now being reported. Sandra Hopps, Michel Pépin, and Jean-Marie Boisvert (2003) evaluated an intervention to decrease loneliness in a small group of people with physical disabilities compared with a small wait-list control group. The intervention was cognitive-behavioral in nature and was administered via a group chat. Loneliness was decreased in the treatment group and effects were maintained at the four-month follow-up. Deborah Tate, Rena Wing, and Richard Winett (2001) randomly assigned overweight adults to a behavioral weight-loss Internet program or to an Internet education program. The Internet treatment resulted in more weight loss and greater reductions in waist measurement at three and six months than the education group. Lasse Strom, Richard Pettersson, and Gerhard Andersson (2000) randomly assigned people with chronic headaches to a six-week Internet-based cognitive behavioral program covering relaxation and problem-solving skills or to a wait-list control group. They found that headache severity was reduced for the treatment group. In fact, 50 percent of those in the treatment group showed clinically significant change compared with only 4 percent in the control group.

Lee Ritterband and colleagues (2003) randomly assigned children with encopresis to an adjunctive treatment group using an interactive psychoeducational and behavioral Web-based program or to a group that did not receive the Web-based treatment. Both groups were encouraged to continue meeting with their physicians. The group who received the Web-based adjunctive treatment improved significantly compared with the control group. The intervention group had fewer fecal accidents, more bowel movements in the toilet, and increased trips to the bathroom compared with the control condition. The results are clearly clinically significant in that the Web-intervention group's number of fecal accidents decreased to one every two weeks, while the control group still experienced more than one accident on average per day.

The most extensive research of mental health interventions using ICTs has been conducted on treatment and prevention for anxiety disorders and eating disorders. For example, Stéphane Bouchard and colleagues (2000) treated eight clients with panic disorder. Face-to-face interviews were conducted first, then the clients received twelve sessions of cognitive-behavioral therapy using videoconferencing technology. Clients used fax machines to submit homework assignments between sessions. A control group was not used in this study, but significant pre-post results were found on all outcome measures despite the small sample.

Britt Klein and Jeffrey Richards (2001) randomly assigned a small sample of clients with panic disorder to a three-week Internet-based program or to a self-monitoring control group. The Internet program was psychoeducational and covered information about methods for managing panic and negative thinking errors. The treatment group had lower panic frequency and general anxiety, and decreased body vigilance after treatment, suggesting that the intervention was effective at least in the short term.

Justin Kenardy, Kelly McCafferty, and Virginia Rosa (2003) have developed the Online Anxiety Prevention Program. This six-week self-administered program is cognitive-behavioral in nature and provides participants with education regarding anxiety, relaxation, cognitive restructuring, interoceptive exposure, and relapse prevention. The researchers randomly assigned undergraduates to the online program or to a wait-list control group. The intervention group had lower anxiety-related cognitions and lower depression scores compared with the control group at the end of treatment.

Alfred Lange and colleagues (2003) have developed an online treatment in The Netherlands called "Interapy" for posttraumatic stress and pathological grief. They randomly assigned clients to a five-week Web-based intervention or to a wait-list control group. The treatment consists of ten forty-five-minute writing assignments (two per week). Therapists reviewed the written work and provided instructions and feedback. The first four writing assignments involved self-exposure: participants wrote about their traumatic event in detail. The next assignments involved cognitive reappraisal: participants wrote letters to hypothetical friends dealing with similar issues. Lastly, participants wrote letters that were not necessarily sent, in which they shared information, came up with a farewell ritual, and discussed how they planned to cope in the future. The intervention led to decreases in intrusions and avoidance symptoms as well as decreased general psychopathology.

Much programmatic work has been done in the area of eating disorder prevention and treatment. In 1994, staff at Wellesley College started an electronic bulletin board so that students could post comments and reply to each other about issues related to body image, food, and eating (Gleason, 1995). The program was not formally evaluated, and there was a lack of confidentiality because all posts on the system automatically identified the writer, but nonetheless many students participated.

Since that time, researchers have developed a structured psychoeducational Web-based program called Student Bodies that has been repeatedly evaluated and adapted. The program is based on cognitive-behavioral body image interventions and uses text, audio, journal entries, and behavioral self-assessments (Low et al., 2006). Angela Celio and colleagues (2000) compared the Student Bodies Internet program combined with limited face-to-face contact to a purely face-to-face group and found that the Internet group did better at reducing body dissatisfaction and disordered eating and attitudes among college women. Subsequent studies dropped the face-to-face component of the Student Bodies program. Andrew Winzelberg and colleagues (2000) found that the Web-based Student Bodies program for college women resulted in increased body satisfaction at the three-month follow-up compared to a control condition, and that the intervention group also had a decreased drive for thinness.

Marion Zabinski and colleagues (2001) randomly assigned a high-risk sample of college women to either the Student Bodies program or to a control group. Both groups were found to improve at the ten-week follow-up on measures of body image and eating behavior. In a later study, Zabinski and colleagues (2004) evaluated a cognitive-behavioral Internet program that is similar to the Student Bodies program that included readings, an asynchronous message board, homework assignments, and weekly synchronous one-hour chats with a moderator. At-risk college students were randomly assigned to one of three ten-person groups or to a thirty-person wait-list control group. The treatment resulted in a reduction of risk factors for eating disorders. The authors indicated that they seemed to have found better results with their synchronous treatment compared to studies using the asynchronous Student Bodies program. However, it should be noted that many participants were screened out of the Zabinski and colleagues (2004) study because they could not make the scheduled chat times. Thus, an asynchronous program is likely easier to implement for large groups.

Kathyrn Low and colleagues (2006) also targeted college women. Participants were randomly assigned to four conditions: the Student Bodies program with a moderated asynchronous discussion group, the Student Bodies program with an unmoderated asynchronous discussion group, the Student Bodies program without a discussion group, and a control group. All treatment conditions did better than the control group, and the outcomes for the intervention groups were not significantly different from one another, indicating that moderation of discussion groups by clinicians may not be a necessary ingredient of the program. The treatment groups had reduced risk factors for eating disorders, and these reductions persisted at the eight- to nine-month follow-up.

Evidence is clearly accumulating indicating that Web-based treatments are more effective than no treatment. In the future, we will likely see an increase in studies directly comparing online services to face-to-face services. However, it has been suggested that if online services are being used by people who would not otherwise seek mental health services, then perhaps online therapy does not have to be as good as or better than traditional face-to-face services, as long as it is better than no treatment (Ritterband et al., 2003).

Advantages of Therapy at a Distance

Perhaps the strongest reasons for conducting therapy using ICTs, as opposed to traditional face-to-face therapy, are convenience and cost. Asynchronous interventions (for example, using email or discussion boards) are convenient because they eliminate the need for scheduled appointments (Suler, 2000). They allow participants to engage in therapy whenever they want, including late at night and on weekends when most traditional therapy options are unavailable. Online interventions eliminate travel time and are accessible for people with physical limitations and for people who otherwise might have difficulty traveling to a therapist's office, such as those with agoraphobia who do not leave their homes. These interventions may also save significant money for people living in rural areas who might otherwise have to travel long distances for services (Schopp, Demiris, & Glueckauf, 2006).

Mental health services online are typically less expensive than traditional services for clients (Manhal-Baugus, 2001), and they also may require less time of clinicians if computer programs or Web sites are used for disseminating or collecting information. A one-time email exchange with a therapist may be free (Griffiths, 2005), and individual online chat sessions with a therapist typically cost around a dollar per minute (Lazlo, Esterman, & Zabko, 1999), which can result in a significant savings compared with the cost of face-to-face therapy. It has also been found that videoconferencing costs are significantly lower than costs associated with client travel or psychological outreach (Schopp, Johnstone, & Merrell, 2000). In a study of online predominantly self-help headache treatment, it took an average of forty minutes per client of therapist time for clients to complete a six-week program plus a four-week postassessment (Strom et al., 2000). Strom and colleagues (2000) indicate that Internet therapy can be twelve times as cost-effective as traditional treatment.

It is unclear whether managed care companies will want to reimburse for therapy conducted using ICTs. While services would probably be less expensive per client, it is likely that more people would utilize online services due to convenience than would otherwise seek services, which could ultimately raise costs for managed care companies (Griffiths, 2005). According to Laura Schopp, George Demiris, and Robert Glueckauf (2006), Medicare started reimbursing in 2001 for treatment conducted with video teleconferencing if clients lived in rural areas or areas where there is a shortage of healthcare professionals. Adrian Skinner and Jason Zack (2004) believe that managed care companies will likely cover online services once their effectiveness has been more clearly demonstrated.

Aside from convenience and affordability, there are many other advantages of conducting mental health interventions using ICTs: frequent email contact, chat sessions, or discussion board use may provide social support and useful information for clients (Glueckauf & Ketterson, 2004). Clients may also feel that online interventions are less stigmatizing and less threatening than traditional treatment (Griffiths, 2001). Furthermore, online interventions are thought to level the power imbalance between a therapist and a client (Murphy & Mitchell, 1998), which is a goal of feminist therapy. Online interventions may also be an initial step toward traditional services. In one consumer satisfaction survey of people receiving e-therapy, it was found that 64 percent receiving e-therapy later tried face-to-face treatment (Metanoia, 1999).

In types of online therapy that are asynchronous, such as email exchanges, therapists can take time to think and formulate responses to clients. (In synchronous types, such as videoconferencing or chat sessions, an immediate response would be expected.) Clinicians can use lag time between email exchanges to consult with supervisors or colleagues before responding (Lange et al., 2003). The written record of contact between therapist and client may also be especially useful for therapists, supervisors, researchers, and clients (Suler, 2000). Clients can review information from previous "sessions" or email without taking up additional therapist time. Moreover, the writing-intensive nature of many online interventions may be particularly therapeutic (Griffiths, 2005). Many researchers have found that writing about emotions, especially emotions related to traumatic experiences, can have positive effects on physical and mental health (Pennebaker & Seagal, 1999; Sheese, Brown, & Graziano, 2004).

Some have theorized that online therapy may result in faster psychological changes than traditional therapy because clients are more apt to get right to the heart of their problems, and they may spend less time focused on nontherapy-related issues (Grohol, 2001). Clients can also access information quickly because many sites typically promise that clients will receive an email response to their questions in twenty-four to forty-eight hours. Schopp and colleagues (2006) also point out that using ICTs can allow for frequent collection and monitoring of client reports, which may allow for earlier detection of problems or relapses.

Disadvantages of Therapy at a Distance

Substantial disadvantages of conducting online interventions also exist. Of utmost concern are ethical and legal issues. It is difficult to ensure complete confidentiality when all information is typed and could be sent out accidentally to others with the click of a button. It is also possible that people could access a therapist's communications if encryption programs are not used. Further problems could stem from clients not having privacy in their own homes or from clients using email addresses that are shared with others. Ethical dilemmas involve being unprepared for emergencies like suicidal ideation if the therapist and client are geographically separated and the therapist is unaware of local resources. It is also possible that clinicians may not recognize when a phone call or a face-to-face session is necessary (Yager, 2001).

Therapy using ICTs also has potentially problematic legal implications, especially related to licensure and credentialing. Clinicians are typically licensed to practice within a given state and thus, if a clinician is conducting therapy via the Internet with a client in another state, the clinician is unlicensed to do so. Clear reciprocity of licensure across state lines does not typically exist, and to get licensed in multiple states is cumbersome and costly. States also vary in the criteria used for licensure. Marlene Maheu and Barry Gordon (2000) did a survey and found that most professionals providing telehealth services were treating people outside of their state. Maxine Capner (2000) raises the interesting question of where the treatment is taking place—at the therapist's location or the client's? Mary Seeman and Bob Seeman (1999) suggest that the client is being served at the therapist's location, much like clients who travel across state lines to receive traditional services. However, it is unclear whether state licensing boards will agree with this assertion.

Web-based therapy also raises the question of how and where clients would file malpractice claims (Nickelson, 1997). Many third-party payors may be unwilling to provide coverage or reimbursement for Web-based services. Clinicians must clearly indicate in insurance forms that services are being provided via the Internet (Koocher & Morray, 2000). Failure to do so could result in fraud charges (Koocher & Morray, 2000). Another important potential legal problem inherent in Web-based therapy concerns the provision of treatment to minors. In many states, parental consent for mental health services is required when a minor presents for treatment. However, minors using ICTs could potentially lie about their age without the clinician's awareness (Griffiths, 2001).

Another potential disadvantage of using ICTs for therapy relates to the impact that the lack of physical presence can have on the relationship that develops between therapist and client. Intimacy may be reduced, and typed text may lead to misunderstandings (Suler, 2000). These problems could be somewhat minimized, however, if videoconferencing technology is used (Suler, 2000), or if clients are told to inform the therapist whenever they find the therapist's communication unsettling (King & Poulos, 1999).

While online interventions using highly structured treatments like many cognitive-behavioral therapies are easy to imagine, it may be far more difficult to engage in less-structured types of therapy or therapy that relies heavily on the therapeutic alliance between client and therapist, such as psychoanalysis (Bouchard et al., 2000). Projections and transference reactions may be exaggerated (Suler, 2000), and problems may occur if a clinician doesn't respond quickly enough to communications from clients (Yager, 2001).

For the therapist using ICTs as an adjunct to traditional services (for example, email between face-to-face sessions), extra clinician time for which the therapist is unlikely to be compensated might be required. Murdoch and Connor-Greene (2000) estimate that adjunctive email may take about ten minutes per client per day. For therapists conducting purely online treatment (for example, chat sessions), treatment might require as much time as traditional services but might yield less payment.

Reducing Legal and Ethical Problems

Many steps for reducing the likelihood of ethical or legal problems resulting from Web-based interventions have already been articulated. For example, Keith Humphreys, Andrew Winzelberg, and Elena Klaw (2000) suggest that clinicians should only treat people who live locally and should screen clients in person before starting Web-based therapy. Some ethical concerns can be alleviated by making online groups and advice "educational" and by making it clear that online groups are public forums and are not to be considered "mental health treatment" (Humphreys et al., 2000). Thus, online groups and their questions and answers could be similar to "talk radio shows" without personal therapeutic relationships (Humphreys et al., 2000).

Alternatively, clinicians could call their services "therapy" but have clients sign a form indicating that they realize that therapy conducted via the Internet is relatively new and untested (King, Engi, & Poulos, 1998). Having an informed consent process for clients starting therapy online should be routine, but only half of all telehealth providers were found to have such a process (Maheu & Gordon, 2000). All clients should be provided with Web site addresses where they can check the certification or licensure status of online clinicians (Bloom, 1998).

Online therapists must have the local addresses of all clients as well as the contact information for local police or crisis centers in the clients' areas in case emergencies arise, such as suicidal ideation or potential child abuse (Childress & Asamen, 1998). Suicidal intentions are of major concern because a clinician conducting therapy using ICTs may not read communications from clients right away (Murdoch & Connor-Greene, 2000). Clients need to be informed about procedures that will be followed in emergencies (Koocher & Morray, 2000), and a contingency plan needs to be in place regarding contact between client and therapist in case of system failure (Skinner & Zack, 2004).

Encryption programs should be routinely used (Childress & Asamen, 1998), and all emails should be printed and placed in a permanent record for each client (Maheu & Gordon, 2000). Furthermore, to ensure confidentiality and to enable self-disclosure, clients need privacy within their homes while using their computer (Sampson, Kolodinsky, & Greeno, 1997), and they need their own personal email address (Yager, 2001). In addition, all copyrighted assessment questionnaires should be mailed to clients' street addresses so that copyright laws are not violated (Childress & Asamen, 1998). Psychologists should let all people in active or large online groups know that the psychologist does not read every post, so that group members do not expect personal responses from the clinician every time they make a post (Humphreys et al., 2000). Also, as previously indicated, if third parties are billed for services, clinicians need to accurately describe on insurance forms that the treatment is occurring online (Koocher & Morray, 2000).

Online Clinical Supervision

Clinical supervision of cases via the Internet is an innovative idea receiving attention in recent years. Clinical supervision is a necessary component of training for new or unlicensed therapists, but sometimes it can be difficult for training programs to find enough supervisors. Jennifer Wood, Thomas Miller, and David Hargrove (2005) suggest that computer-mediated supervision models can increase exposure to supervision and exposure to certain types of expertise, especially in rural areas. They employed telesupervision using Web sites, email, and videoconferencing as an adjunct to group face-to-face supervision. Deede Gammon and colleagues (1998) also describe a hybrid method of supervision that combines face-to-face and online supervision. Supervisors initially meet face-to-face and establish a relationship with the trainees, then the trainees complete up to half of their supervision hours via videoconferencing. Dean Janoff and Judith Schoenholtz-Read (1999) also blend online supervision with face-to-face meetings. Their supervision program at the Fielding Institute entails eight seven-hour face-to-face group supervision meetings per year. These meetings occur on weekends roughly every six weeks. In between these meetings, supervisors and trainees stay in contact via email and threaded discussions on the Fielding Web site. No identifying information about clients is ever transmitted online, and all clients sign a consent form indicating that they realize that their clinician will be receiving group supervision in person and online.

Online supervision and face-to-face supervision of cases in which the therapy is conducted online are unique in that the supervisor can have a transcript of the therapy sessions and has as much access to the client as the trainee does (Murphy & Mitchell, 1998). Also, trainees conducting online therapy can review their potential responses to patient communication with their supervisors before sending them (Murphy & Mitchell, 1998). While online supervision hours may work well and may alleviate some problems for training programs that are having difficulty finding available supervisors, it should be noted that some states require that trainees count only face-to-face supervision hours for licensure requirements (Wood et al., 2005). Whether this requirement will be changed due to the needs of training programs is yet to be seen.

Conclusion

Therapy at a distance is being used to treat a variety of mental health problems, including anxiety disorders, eating and weight issues, loneliness, headaches, pediatric traumatic brain injury, child behavior problems, pediatric encopresis, and problems related to caring for people with schizophrenia and Alzheimer's disease. The most programmatic work is being done in the areas of anxiety disorders and eating/weight disturbances. Evidence is starting to accumulate suggesting that mental health interventions using ICTs are more beneficial than no treatment. Furthermore, it seems that people who might not otherwise seek traditional face-to-face services are seeking these interventions. Therapy and clinical supervision using ICTs are often more convenient and less expensive than traditional services. However, legal and ethical challenges are inherent in the work. Despite these challenges, it is expected that therapy and supervision using ICTs will gain in popularity.

References

Abascal, L., Brown, J., Winzelberg, A. J., Dev, P., & Taylor, C. B. (2004). Combining universal and targeted prevention for school-based eating disorder programs. *International Journal of Eating Disorders*, 35, 1–9.

Bloom, J. W. (1998). The ethical practice of webcounseling. *British Journal of Guidance and Counselling*, 26(1), 53–59.

Bouchard, S., Payeur, R., Rivard, V., Allard, M., Paquin, B., & Renaud, P. (2000). Cognitive behavior therapy for panic disorder with agoraphobia in videoconference: Preliminary results. *CyberPsychology and Behavior*, 3, 999–1007.

Capner, M. (2000). Videoconferencing in the provision of psychological services at a distance. *Journal of Telemedicine and Telecare*, 6, 311–319.

Castelnuovo, G., Gaggioli, A., Mantovani, F., & Riva, G. (2003). New and old tools in psychotherapy: The use of technology for the integration of traditional clinical treatments. *Psychotherapy: Theory, Research, Practice, Training*, 40(1–2), 33–44.

Celio, A. A., Winzelberg, A. J., Taylor, C. B., Wilfley, D. E., Eppstein-Herald, D., Springer, E. A., & Dev, P. (2000). Reducing risk factors for eating disorders: Comparison of an Internet- and a classroom-delivered psychoeducational program. *Journal of Consulting and Clinical Psychology*, 68, 650–657.

Childress, C. A., & Asamen, J. K. (1998). The emerging relationship of psychology and the Internet: Proposed guidelines for conducting Internet intervention research. *Ethics and Behavior*, 8, 20–28.

Gammon, D., Sorlie, T., Bergvik, S., & Hoifodt, T. S. (1998). Psychotherapy supervision conducted by videoconferencing: A qualitative study of users' experiences. *Journal of Telemedicine and Telecare*, 4, 33–35.

Gleason, N. A. (1995). A new approach to disordered eating using an electronic bulletin board to confront social pressure on body image. *Journal of American College Health*, 44, 78–80.

Glueckauf, R. L., & Ketterson, T. U. (2004). Telehealth interventions for individuals with chronic illnesses: Research review and implications for practice. *Professional Psychology: Research and Practice*, 35, 615–627.

Glueckauf, R. L., Stine, C., Young, M. E., Bourgeois, M., Pomidor, A., Rom, P., Massey, A., & Ashley, P. (2005). Alzheimer's rural care healthline: Linking rural caregivers to cognitive-behavioral interventions for depression. *Rehabilitation Psychology*, 50(4), 346–354.

Griffiths, M. D. (2001). Online therapy: A cause for concern? *The Psychologist*, 14, 244–248.

———. (2005). Online therapy for addictive behaviors. *Cyberpsychology and Behavior*, 8(6), 555–561.

Grohol, J. M. (2001). Best practices in e-therapy: Clarifying the definition of e-therapy. Retrieved May 24, 2006, from http://psychcentral.com/best/best5.htm.

Hopps, S. L., Pépin, M., & Boisvert, J. (2003). The effectiveness of cognitive-behavioral group therapy for loneliness via inter-relay-chat among people with physical disabilities. *Psychotherapy: Theory, Research, Practice, Training*, 40(1–2), 136–147.

Hornblow, A. R. (1986). The evolution and the effectiveness of telephone counseling services. *Hospital and Community Psychiatry*, 37, 731–733.

Humphreys, K., Winzelberg, A., & Klaw, E. (2000). Psychologists' ethical responsibilities in Internet-based groups: Issues, strategies, and a call for dialogue. *Professional Psychology: Research and Practice*, 31, 493–496.

Janoff, D. S., & Schoenholtz-Read, J. (1999). Group supervision meets technology: A model for computer-mediated training at a distance. *International Journal of Group Psychotherapy*, 49(2), 255–272.

Jerome, L. W., DeLeon, P. H., James, L. C., Folen, R., Earles, J., & Gedney, J. J. (2000). The coming age of telecommunications in psychological research and practice. *American Psychologist*, 55, 407–421.

Kenardy, J., & Adams, C. (1993). Computers in cognitive behaviour therapy. *Australian Psychologist*, 28, 189–194.

Kenardy, J., McCafferty, K., & Rosa, V. (2003). Internet-delivered indicated prevention for anxiety disorders: A randomized controlled trial. *Behavioural and Cognitive Psychotherapy*, 31, 279–289.

King, S. A., Engi, S., & Poulos, S. T. (1998). Using the Internet to assist family therapy. *British Journal of Guidance and Counselling*, 26(1), 43–52.

King, S. A., & Poulos, S. T. (1999). Ethical guidelines for on-line therapy. In J. Fink (Ed.), *How to Use Computers and Cyberspace in the Clinical Practice of Psychotherapy* (pp. 121–132). Blue Ridge Summit, PA: Jason Aronson Publishers.

Klein, B., & Richards, J. C. (2001). A brief Internet-based treatment for panic disorder. *Behavioural and Cognitive Psychotherapy*, 29, 113–117.

Koocher, G. P., & Morray, E. (2000). Regulation of telepsychology: A survey of state attorneys general. *Professional Psychology: Research and Practice*, 31, 503–508.

Lange, A., Rietdijk, D., Hudcovicova, M., van de Ven, J., Schrieken, B., & Emmelkamp, P. M. G. (2003). Interapy: A controlled randomized trial of the standardized treatment of post-traumatic stress through the Internet. *Journal of Consulting and Clinical Psychology*, 71(5), 901–909.

Lazlo, J. V., Esterman, G., & Zabko, S. (1999). Therapy over the Internet? Theory, research, and finances. *Cyberpsychology and Behavior*, 2(4), 293–307.

Low, K. G., Charanasomboon, S., Lesser, J., Reinhalter, K., Martin, R., Jones, H., et al. (2006). Effectiveness of a computer-based interactive eating disorders prevention program at long-term follow-up. *Eating Disorders*, 14, 17–30.

Luce, K. H., Winzelberg, A. J., Osborne, M. I., & Zabinski, M. F. (2003). Internet-delivered psychological interventions for body image dissatisfaction and disordered eating. *Psychotherapy: Theory, Research, Practice, Training*, 40(1–2), 148–154.

Maheu, M. M., & Gordon, B. L. (2000). Counseling and therapy on the Internet. *Professional Psychology: Research and Practice*, 31, 484–489.

Manhal-Baugus, M. (2001). E-therapy: Practical, ethical and legal issues. *CyberPsychology and Behavior*, 4, 551–563.

Metanoia. (1999). ABC's of Internet therapy: Survey results. Retrieved August 27, 2006, from http://www.metanoia.org/imhs/survey.htm.

Morawska, A., & Sanders, M. R. (2006). Behavioral family intervention for parents of toddlers, part 1: Efficacy. *Journal of Consulting and Clinical Psychology*, 74(1), 10–19.

Murdoch, J. W., & Connor-Greene, P. A. (2000). Enhancing therapeutic impact and therapeutic alliance through electronic mail homework assignments. *Journal of Psychotherapy: Practice and Research*, 9, 232–237.

Murphy, M. J. (2003). Computer technology for office-based psychological practice: Applications and factors affecting adoption. *Psychotherapy: Theory, Research, Practice, Training*, 40(1–2), 10–19.

Murphy, L. J., & Mitchell, D. L. (1998). When writing helps to heal: Email as therapy. *British Journal of Guidance and Counselling*, 26(1), 21–32.

Newman, M. G. (2000). Recommendations for a cost-offset model of psychotherapy allocation using generalized anxiety disorder as an example. *Journal of Consulting and Clinical Psychology*, 68(4), 549–555.

Nickelson, D. (1997, August). Telehealth poses opportunities and challenges for psychology. *Practitioner Focus* (newsletter of the American Psychological Association, Public Relations and Communications Practice Directorate). Retrieved March 8, 2006, from http://www.apa.org/practice/pf/aug97/teleheal.html.

———. (1998). Telehealth and the evolving health care system: Strategic opportunities for professional psychology. *Professional Psychology: Research and Practice*, 29, 527–535.

Pennebaker, J. W., & Seagal, J. D. (1999). Forming a story: The health benefits of narrative. *Journal of Clinical Psychology*, 55, 1243–1254.

Postel, M. G., de Jong, C. A. J., & de Haan, H. A. (2005). Does e-therapy for problem drinking reach hidden populations? *American Journal of Psychiatry*, 162(12), 2393.

Ritterband, L. M., Cox, D. J., Walker, L. S., Kovatchev, B., McKnight, L., Patel, K., Borowitz, S., & Sutphen, J. (2003). An Internet intervention as adjunctive therapy for pediatric encopresis. *Journal of Consulting and Clinical Psychology*, 71(5), 910–917.

Rosenfield, M., & Smillie, E. (1998). Group counselling by telephone. *British Journal of Guidance and Counselling*, 26(1), 11–19.

Rotundi, A. J., Haas, G. L., Anderson, C. M., Gardner, W. B., Newhill, C. E., Spring, M. B., et al. (2005). A clinical trial to test the feasibility of a telehealth psychoeducational intervention for persons with schizophrenia and their families: Intervention and three-month findings. *Rehabilitation Psychology*, 50(4), 325–336.

Sampson, J. P., Kolodinsky, R. W., & Greeno, B. P. (1997). Counseling on the information highway: Future possibilities and potential problems. *Journal of Counseling and Development*, 75, 203–212.

Sansone, R. A. (2001). Patient-to-patient e-mail: Support for clinical practices. *Eating Disorders*, 9, 373–375.

Schopp, L., Demiris, G., & Glueckauf, R. L. (2006). Rural backwaters or front-runners? Rural telehealth in the vanguard of psychology practice. *Professional Psychology: Research and Practice*, 37(2), 165–173.

Schopp, L., Johnstone, B., & Merrell, D. (2000). Telehealth and neuropsychological assessment: New opportunities for psychologists. *Professional Psychology: Research and Practice*, 31, 179–183.

Seeman, M. V., & Seeman, B. (1999). E-psychiatry: The patient-psychiatrist relationship in the electronic age. *Canadian Medical Association Journal*, 161, 1147–1149.

Sheese, B. E., Brown, E. L., & Graziano, W. G. (2004). Emotional expression in cyberspace: Searching for moderators of the Pennebaker disclosure effect via e-mail. *Health Psychology*, 23(5), 457–464.

Skinner, A., & Zack, J. S. (2004). Counseling and the Internet. *American Behavioral Scientist*, 48(4), 434–446.

Strom, L., Pettersson, R., & Andersson, R. (2000). A controlled trial of self-help treatment of recurrent headache conducted via the Internet. *Journal of Consulting and Clinical Psychology*, 68(4), 722–727.

Suler, J. R. (2000). Psychotherapy in cyberspace: A five-dimensional model of online and computer-mediated psychotherapy. *CyberPsychology and Behavior*, 3, 151–159.

Tate, D. F., Wing, R. R., & Winett, R. A. (2001). Using Internet technology to deliver a behavioral weight loss program. *Journal of the American Medical Association*, 285, 1172–1177.

Wade, S. L., Wolfe, C. R., Brown, T. M., & Pestian, J. P. (2005). Can a Web-based family problem-solving intervention work for children with traumatic brain injury? *Rehabilitation Psychology*, 50(4), 337–345.

Winzelberg, A. J., Eppstein, D., Eldredge, K. L., Wilfley, D. E., Dasmahapatra, R., Dev, P., & Taylor, C. B. (2000). Effectiveness of an Internet-based program for reducing risk factors for eating disorders. *Journal of Consulting and Clinical Psychology*, 68, 346–350.

Wood, J. A. V., Miller, T. W., & Hargrove, D. S. (2005). Clinical supervision in rural settings: A telehealth model. *Professional Psychology: Research and Practice*, 36(2), 173–179.

Yager, J. (2001). E-mail as a therapeutic adjunct in the outpatient treatment of anorexia nervosa: Illustrative case material and discussion of the issues. *International Journal of Eating Disorders*, 29, 125–138.

Zabinski, M., Pung, M. A., Wilfley, D. E., Eppstein, D. L., Winzelberg, A. J., Celio, A., & Taylor, C. B. (2001). Reducing risk factors for eating disorders: Targeting at-risk women with a computerized psychoeducational program. *International Journal of Eating Disorders*, 29, 401–408.

Zabinski, M. F., Wilfley, D. E., Calfas, K. J., Winzelberg, A. J., & Taylor, C. B. (2004). An interactive psychoeducational intervention for women at risk of developing an eating disorder. *Journal of Consulting and Clinical Psychology*, 72(5), 914–919.

Zhu, S., Tedeschi, G. J., Anderson, C. M., & Pierce, J. P. (1996). Telephone counseling for smoking cessation: What's in a call? *Journal of Counseling and Development*, 75(2), 93–102.

· 13 ·

BUT YOU DON'T PLAY WITH THE MOBILE INFORMATION AND COMMUNICATION TECHNOLOGIES YOU ALREADY HAVE

An Instructional Technologist's View of Teaching with Technology in Higher Education

Gary Pandolfi

When my clients hear the word *technologist* in my job title, they believe that I can solve every problem they have with their computer hardware, software, or Internet access provider. While I do help when I can, my thirty years of teaching and instructional design experience in the classroom and in industry make me more useful as a consultant. My reply to "What does an instructional technologist do?" is that I provide faculty with a variety of opportunities to incorporate technology into their teaching in ways that can make teaching more effective, and learning more efficient and more significant. In this chapter, I discuss several examples of teaching with mobile information and communication technologies (ICTs).

I love technology. I embrace it wholeheartedly. Years ago, as a middle-school English teacher, I was the first at my school to figure out how to make my student progress reports fit the triplicate dot-matrix forms we were compelled to use to record our comments at the end of each marking period. My 1995 presentation to the English department at Kingswood-Oxford School, "Exemplary Baroque Pieces," projected a transparency of Antonio Pereda's *Knight's Dream* created with a color scanner, color printer, and black-and-white PowerBook, while the attached CD-Rom drive played Vivaldi's *Concerto in Do Maggiore per Mandolino* through external speakers. I employed databases and drawing programs as I designed online learning activities to complement McGraw-Hill textbooks. And today my college-level English class is virtually paperless because students post their papers and I correct them electronically through a Web-based course management system. Technology is definitely my friend.

So how do I respond to the professor who has read about a class at another institution that made "podcasts" of conversations about artwork in the Metropolitan Museum of Modern Art and then published them on the World Wide Web and wants to "do that"? (See Hanson and Baldwin, chapter 8 in this volume, for a discussion of podcasting.) Actually, I balk. Well, I don't balk, exactly—I begin asking questions:

1. Why do you want to use an iPod?
2. Will iPods make something you do now easier? How?
3. What are the disadvantages of the teaching model you currently use?
4. Will you use this new technology to support a particular learning objective?
5. Will this objective replace one that you already have or will it add something to the course?
6. Are there music or video files that students will access using this device?
7. Is there any other electronic device they currently use that will do the same thing?

Although the word *technologist* in my title suggests to my colleagues that I know everything about technology and can make anything they request happen immediately, I see my role as more like that of a cooking show host who asks questions of an accomplished chef in order to help the audience understand what is being done. Instead of asking "What are you making?" I ask, "What are your learning objectives?" Instead of "What ingredients are you using?" I ask, "What methods do you use?" "Why did you use powdered vanilla instead of liquid?" becomes "Why do you prefer the lecture format over a more constructivist problem-solving approach?" These questions become increasingly important when the teacher is redesigning a course. "Have you ever used online formative assessments before?" "What low-stakes writing assignments do you use?"

Like the chef, professors begin to anticipate my questions and explain as they add materials, or ask for materials to be made available in certain ways. "I want to digitize this lecture so students can review it later and so that they won't be trying to write down everything I say in class." The professors become reflective of their own processes, making their decisions more deliberate and careful. "I considered using an online quiz, but then I decided that they should take a quiz during the first five minutes of class. This ensures that they are doing their own work." This reflective habit of mind allows professors to try new things with minimal risk of failure and maximum value, even if they decide not to repeat the exercise the same way next time.

In cooking shows, the chefs are culinary experts, as our professors are experts in their own disciplines. However, professors are often not expert teachers, because they are not subject to rigorous certification standards that pertain to public secondary and elementary teaching assignments. Many try to teach the way their professors taught them, often not taking into account today's student population and current findings in learning theory research.

Moreover, professors often see a new technology demonstrated in an engaging way and want to adopt it immediately, unaware that the same results could be obtained from technology that is already in place at the institution. While I don't say, "But you don't play with the technology we already have in place," I do suggest that they experiment with what we have first. This practice allows them to identify shortcomings of the present software or hardware, and to demonstrate specific advantages of a new technology. In other words, by vetting their methodology with the technology currently available, they can make a stronger case for new technology that makes what they are doing more accessible.

For example, a professor uses email for asynchronous class discussion. He emails the entire class with a question and instructs them to use the "reply to all" feature. He instructs them to post a minimum of one reply to the question, and one reply to someone else's reply. He has great success with this process. When this professor hears about a software package that will allow him to have small group or entire class discussions within his course management system, he is in a better position to lobby for the advantages of this new technology over his current practice.

Let us suppose that the previously mentioned professor who wanted to "do that" with podcasts is teaching an introductory art course, the overarching objective of which is that by the end of the course students will have demonstrated an understanding of and appreciation for art. Teaching students audio recording, editing, and Web publishing so that they can record a live conversation about a few paintings at the Metropolitan Museum of Art and then publish them for review and evaluation might take the focus away from the course objective. This assignment could help students develop an appreciation for and understanding of audio production and allow them to demonstrate those qualities, but that is quite a different objective, perhaps more suited to a communications course. If, however, the podcasting technology had a relatively shallow learning curve, and assessment focused on the students' discussion of the art, rather than on how well the students used the technology, it might be a great way for students to demonstrate mastery of the course objective.

The project could also be done twice. Students could make one presentation using PowerPoint that discusses certain art pieces that reside at a local museum. They could find the images on the Web to create slides and use the recording feature to capture a discussion of the images. Then they could visit the actual art pieces and repeat the exercise to highlight the difference between seeing an image online and seeing the actual work. The project could have an added bonus if sound files of the students' presentations are used as learning objects to support the course in the future. Faculty could post these sound files in a course management system and ask students to evaluate the conversations as a formative exercise. In this case, developing a teaching model that uses this technology could be worthwhile. Class time could now be used to discuss presentations students viewed outside of class, facilitated by the expert, the professor.

Now that we have a plan, what technology do the students have in hand for accomplishing this project? Because our university has a laptop requirement for all undergraduates, students can use these devices to capture their discussions. Using an inexpensive headset with attached microphone or even the built-in microphone that some laptop models have, they can fulfill the requirements of the assignment. They can save their presentations to CD, or they can publish them in their university-provided Web space. They can present them to the professor, who can then post them on the course management system for the entire class. Students can download the files and review them at their leisure while traveling or relaxing.

The professor must be careful to evaluate the presentations based on the course objectives and not on the quality of the recording. Did the discussion of the art use appropriate terminology? Was the student's evaluation thorough? Any criteria that would be considered in a written work can apply here. And the entire project can be accomplished with existing technology. The results demonstrate mastery of the course material. Using an iPod becomes unnecessary, as does podcasting. Moreover, my cooking show host methodology prepares the faculty member to "own" the project, by which I mean that he or she understands its advantages and disadvantages, adding to the likelihood that it will succeed.

Technology can help teachers and students make better use of the time they have together and ensure that students get the education they expect. Students attend institutions of higher learning for the opportunity to benefit from their professors' expertise. They are not paying just for content. Certainly they need guidance as to what content to absorb, what information to gather, but what they really want to learn is how the content is considered by experts in the field. If professors can shift the responsibility for learning content onto the students outside of class and spend the class time helping students to make sense of it, students are getting the value they expect.

When I asked a group of art professors, "How many of you feel that you have enough class time?" no one raised a hand. One said that she assigns reading that students don't do. When they come to class, she found herself having to provide the content covered in the reading assignment. The result was that those who read the assignment felt that they had wasted time on the reading and so neglected to do it the next time. I certainly wouldn't struggle to understand a text if the professor was going to distill it for me at the next class. Although the memory of reading quizzes I gave my middle-school students to make sure they had done the reading flashed briefly to mind, I considered that this is higher education. Students have to be treated as adults or nearly adults. I suggested that it might be helpful to post questions or have students post questions derived from the reading in a discussion board. In this way she could tailor her class to explore areas that were less clear to her students. I proposed that she create short, formative, low-stakes assessments online to help students, and the teacher, discover what the students understood from the reading in advance of class. Then class discussion could target those areas that students found difficult while reinforcing that students are responsible for the content of the reading. I suggested that rather than lecturing during each class meeting, the professor could post in the course management system a couple of images that were not in the text that students should be able to say something about based on their reading. Another approach would be to have the students find additional images to support the reading content. The results would trigger engagement with the material under the facilitation of the expert professor. It seems to me that these kinds of activities result in more significant learning outcomes than just memorized content.

Understanding that professors have to "own" the technology first, become completely comfortable with it and understand its advantages and disadvantages, there is much to be said for technologies that make learning more efficient and more accessible to students. With a workload expectation of three hours outside of class for every hour in class, students need to be efficient learners. Technologies that allow students to make better use of their otherwise unscheduled time fall into this category.

Consider the portable lecture made possible with audio recording freeware and digital media players. Students can download the lecture and listen to it on a train, while waiting for a bus, while hiking a local trail. Is this better than sitting through a one-hour, one-shot presentation in a lecture hall packed with a hundred other students? Sometimes. Might an individual benefit from being able to rewind and play back sections of the lecture? Possibly. Would it be easier to take notes in such a format? The main point is that such practices provide students with choices. By creating more opportunities for students to obtain course content, by creating multiple methods for content delivery, we can legitimately hold all students responsible for that content, no matter what their preferred learning style might be.

What are the losses? Portable lectures offer a poor substitute for a dynamic lecturer who captivates students with body language, hand gestures, spontaneous examples, and changes in pitch or volume, while engaging students in some work at various intervals. It's like listening to a CD instead of the live performance. Making the audio file available after the lecture for study and review purposes may also cause students to become passive during the actual lecture. Students may opt to skip the lecture, preferring to wait until it comes out in MP3 (a digital audio file). Students may see these recordings as a shortcut, a substitute for hard work.

Despite these legitimate concerns, enabling students to engage the ideas in the lecture during class time and then review them at their leisure seems to be beneficial. In order for these technologies to be effective, students must be educated on their use as an additional resource rather than as a shortcut. First, students must understand that there is no substitute for hard work. This means that professors need to keep standards high while providing every opportunity possible for students to achieve those standards. Students must be educated to write questions and observations of their own in their notes rather than transcribe content. They should annotate a lecture as they might annotate a text. When I was a student, the only way to preserve the lecture content was to transcribe as much of it as possible, or request permission to tape the lecture, a poor substitute in a visual-dependent lecture such as my Architecture 101 class.

Early in my tenure here at Quinnipiac University, an adjunct biology professor wanted to save his PowerPoint presentation to a CD for a student whose ill health caused him to miss a lecture. I told the professor that we could certainly do that, and then I asked, "Did you know that we can post this PowerPoint presentation into our course management system? Then all your students could view it. They would not have to try to copy down everything you say, but they could write what they think about the lecture and questions they may have. Wouldn't this benefit everyone?"

"But if my lectures are online, why would students come to class?" he asked.

"I'm sure there's more to your lectures than just reading the slides," I said. Then I explained that he could post the slides and arrange for them to become accessible for student viewing and downloading after the actual lecture. This would allow all students to record what they were thinking about during the lecture rather than trying to transcribe the professor's words. They would not have to worry about how fast they could write or whether they missed anything. The professor agreed that posting his presentations could benefit all students.

An even greater improvement would result from the professor not lecturing in class, instead using class time in other, more engaging, student-centered ways. He could, for example, create a cooperative activity in which students used the information from the lecture to solve a hypothetical problem. Would this approach result in more significant learning?

During a workshop introducing new users to our course management system, an anatomy lecturer said that his was a very tedious course, but students needed to know the content. The content, I assumed, was in the text, and also in a software program that allowed students to virtually view the human body in layers that revealed every type of tissue. I suggested that he could ask students questions such as "What would happen if this particular bone were much larger? How would you modify the structure of the arm so that human beings could scratch their entire backs? Would this modification cause any complications? What other body parts would this change influence? Might there be additional advantages to such a modification?" Discussing these kinds of problems in class would engage students with the material and help them make it their own.

It would certainly benefit some students to be able to listen to a lecture at a time most conducive to their own learning idiosyncrasies. Cell phones and digital media players can now access presentations that include voice and images. The students could take notes, replaying difficult parts of the lecture to make sense of it at their own pace. They could match parts of the text to parts of the lecture. They could listen more actively, writing down their thoughts rather than transcribing the professor's words. Of course, for the professor the workload initially increases, as he must first record audio with the PowerPoint or other presentation software, such as Macromedia Breeze, and then create robust, engaging class activities that foster understanding of the lecture and reading content. However, such an active application of the material in teacher-facilitated activities allows for a deeper understanding of course concepts and accommodates multiple learning styles.

As new technologies emerge, the possibilities that they may help solve situational factors we all experience in our teaching seem endless, and there are wonderful ways that these new technologies can enhance the learning experiences of our students. However, it is essential to vet the technology in sound teaching practices and to educate our students on how best to take advantage of these tools in order to foster significant learning.

• 14 •

PUMPING UP THE PACE

The Wireless Newsroom

Andrew Smith

It would be impossible for the situation I found myself in on June 29, 1993, to ever happen again.

It was a pleasantly warm, early summer day on Long Island, where I was a reporter for *Newsday*, the daily newspaper in New York City's oldest suburbs. The day before, a landscaper named Joel Rifkin had been pulled over in his Mazda pickup truck after police noticed it had no license plates. In the back was a woman's rotting corpse. During an interrogation, Rifkin told police that this was the seventeenth woman he had killed in the past four years. It was a big story, and my piece of it was to go to the unassuming neighborhood in East Meadow, New York, where Rifkin lived with his mother. Like dozens of reporters from other news organizations, I was to talk to neighbors and, more important, I was to talk to Rifkin's mother or his sister, both of whom were cloistered inside the home where Rifkin had dismembered many of his victims. And, of course, I was to stay in touch with the assigning desk.

It was a typical New York–area media scene. Television trucks parked on the block; crew members took out lawn chairs, ready for a long wait. Reporters made their way door-to-door, talking to neighbors eager to share platitudes about Rifkin. And every ten or fifteen minutes, some reporter would dutifully ring the Rifkins' bell, wait in vain for someone to come to the door, and retreat.

This being 1993, the hardest part of the assignment was staying in touch with the desk. These were the dark days before cell phones, before BlackBerry devices, before laptop computers with wireless access. The only way I could contact the desk was by payphone, and the nearest one was at a convenience store more than a half mile away. Contacting the desk meant leaving the scene.

But contacting the desk was crucial. Editors needed to know what I had and didn't have, so they could decide where other reporters should be sent. I also needed to periodically dump my notes—that is, dictate notes of my interviews—to a reporter in the office, so the story could be written more efficiently when deadline bore down on us. So every few hours—or more often, if I got paged—I'd have to risk something happening at the scene to drive to the payphone and call in to the desk.

It should not be hard to predict what happened when I got paged in the middle of that day. I was gone for no more than fifteen minutes, but those were the fifteen minutes when Rifkin's sister decided it was time to walk the dog. I heard later that she did not say much to reporters, but I missed her words, her tears, her taut expression, and the most frazzled walk that dog probably ever had. I was mortified, happy only that I wasn't in the newsroom to hear the editors react.

Such a screw-up would be absurd now.

Since the dawn of journalism, all the best reporting has required getting out of the newsroom and into the world—City Hall, union halls, crime scenes, war zones, and sports stadiums. But despite the steady march of communication technology, the mechanics of gathering information and getting it into print or onto the air remained cumbersome for decades. Reporters in the field who needed background information had to call their libraries or editors to check facts or find new sources. When it came time to file a story on deadline from an ongoing news event, reporters had to compose in their heads and "write" on the phone with the help of rewrite people at the other end.

The arrival of relatively affordable cell phones in the past decade changed the nature of news reporting more radically than anything since the telegraph, which made possible same-week coverage of the Civil War. Now, a reporter did not have to leave the scene of breaking news to find a payphone, whether it was to do other interviews or to call an editor. Wireless email devices made it even easier to conduct research on the run and allowed reporters to file and update stories while they were still happening.

"When I was a reporter for the *Boston Herald* [from 1987 to 1991], editors dispatched us from the office with bulky, handheld walkie-talkies that had long, immobile antennae, a limited range, and no car charge capability," says Gary Witherspoon, now an assistant metro editor at the *Boston Globe*.

> We couldn't exchange vital information over the radio, lest it be heard by a rival shop. The editor would tell the reporter to find a landline, which made communication a tad tedious. Pagers weren't much better, as reporters still had to find payphones. Wireless communication, I think, has helped streamline newsgathering. Reporters can now feed from the scene, which makes for fresher, more vivid journalism. (G. Witherspoon, personal communication, August 7, 2006)

News can also be more timely—a crucial factor now that newspapers compete in real time with other news outlets on the Internet. Earlier this decade, when I covered criminal courts, my BlackBerry both made my life easier and gave me a jump on competitors. For the first time, I didn't have to step out of the courtroom and into the hall to make a phone call to my editor and tell her what had just happened. I could send an email from my seat and not miss a thing. Even more important, on big stories I could write up-to-the-minute updates for *Newsday*'s Web site, and not have to wait for a break in the action or have to waste another reporter's time dictating my update. The courtroom had become the newsroom.

Wireless devices not only extend the newsroom, they also extend the news-gathering day. In a wireless world, it's not just journalists with BlackBerry email devices and wireless laptops—it's news sources, too. That improves the quality of stories for readers. Reporters less often have to say that a newsmaker could not be reached for comment on a late-breaking story. When sources can be found through their cell phones and BlackBerries, it makes it somewhat "more likely that one can get comment after hours or on Sundays," according to *Newsday* reporter Jennifer Smith (personal communication, August 7, 2006).

Where wireless communication devices have changed reporting the most is where reporting is most frenzied—in war zones and at the scenes of disasters. These are the places where it is most difficult to find information that brings context to chaos, and where it is most challenging to communicate with editors and transmit stories. Now, thanks in particular to widespread wireless Internet connections, there is no excuse for incomplete reporting from the field.

"Without wireless devices, I couldn't do my job," says Matthew McAllester, a foreign correspondent for *Newsday* who has covered wars in the Balkans and the Middle East. When he first started doing this work in the 1990s, "the choice was to dictate the story via a landline or to use a Tandy portable computer (I hesitate to call them laptops) that you could hook up directly to the phone line in your hotel room and feed the text, slowly and directly, into a *Newsday* computer. Both processes were slow and error-prone. It's easier now. 'Do you have Wi-Fi?' is about the first question I ask when booking into a hotel" (M. McAllester, personal communication, August 8, 2006).

Thanks to wireless access, war correspondents now have easy access to what government officials or rebel groups are saying online. Even covering elections is easier. When McAllester covered Montenegro's vote to become independent of Serbia, he was able to talk with Montenegrans in the street and then check election results online while writing the story from a café equipped as a wireless hot spot.

In addition to a satellite phone and a satellite data transmission device, McAllester also travels with a wireless laptop and two cell phones.

> The reason for the two cell phones: In one I keep a SIM card from my British mobile company and in the other I will often use a local SIM card, which I have purchased from the Palestinian, Bosnian or Lebanese phone company, depending on the country [in which I am working]. My British phone is quad-band—a Motorola Razr—and works on nearly every GSM network in the world, enabling me to always have a phone line if I'm in a GSM network area. The calls are expensive, though, and so that's why I have the local number. On the local phone I can make and receive calls cheaply, using credit purchased from scratch cards at local stores. Having a local number makes me a lot more accessible to local people, who would balk at the cost of calling a British cell phone. (M. McAllester, personal communication, August 8, 2006)

All those gadgets run on batteries, and there is not always electricity handy to recharge them. "So I also carry power cables that can plug into the cigarette lighter in a car or crocodile clips that can attach to a car battery. While all of these gadgets can weigh you down a bit, there's no question that they save huge amounts of time, freeing you up to do more reporting and writing. Gone are the days—days I never experienced, thank God—of visiting the telex office late at night to file stories" (M. McAllester, personal communication, August 8, 2006).

Modern-day coverage of disasters has been transformed as well. After Hurricane Katrina destroyed parts of Louisiana and Mississippi, it would have been a nightmare to try to report and write stories from there without wireless communication. Normal phone service had been nearly wiped out; the nearest reliable landlines were far from where the news was unfolding. In particular, air cards—credit card–sized devices that plug into a laptop's PCMCIA slot and allow it to make broadband connections through cell phone towers—made life easier for reporters and photographers.

"The ability to use so-called air cards to hook up to the Web and transmit stories while in the field after Katrina was amazing," says Alex Martin, who spent weeks in New Orleans for *Newsday* after the storm (personal communication, August 7, 2006). "I used it to find directions, to check facts, and to file stories and pictures. I would have had to drive miles out of my way each day to unreliable phone lines to do that, seriously hampering reporting."

Because he did not have to do that, Martin had hours every day that he wouldn't have had otherwise to find stories, do research, meet people he would write about—or just relax for a little bit in a stressful, nonstop environment.

Katie Thomas, another *Newsday* reporter who covered Katrina, says that the contrast could be jarring: "We were looking at this apocalyptic scene—chaos—and we had high-speed Internet access. Anywhere a cell phone worked, we could send our stories" (personal communication, August 7, 2006). In addition, Thomas says she had access to email and the Web, which allowed her to monitor what other news organizations were doing and to adjust if necessary.

Without wireless access, Thomas not only would have spent hours dictating her stories over the phone, but the time of another reporter on Long Island would have been tied up transcribing what Thomas dictated.

Mobile communication technologies are changing the way traditional wire services cover news too. Wire service reporters knew the twenty-four-hour news cycle long before anyone else did; they filed constant updates to their stories. Until recently, they lived the stereotype of news reporters rushing out of a courtroom or a news conference, frantically calling an editor, demanding a rewrite person, and dictating a new version of an evolving story. But no more.

Now, with laptops and air cards, wire service reporters in the field are their own rewrite people; they take laptops to the scene of breaking news and write their stories from there. And wire services now can use their resources more intelligently as a result. There are more reporters in the field—where the news happens—and fewer rewrite people waiting by the phone for reporters to call in. As Witherspoon notes, it is a situation that makes for more vivid, more thorough reporting.

Months after Katrina, when Thomas returned to New Orleans to do research for her book, *Waters Dark and Deep: How One Family Overcame Hurricane Katrina's Deadly Fury* (2006), she brought her laptop. With wireless access, she could verify facts and get maps she needed, speeding her work there.

For photographers, the advent of wireless mobile communication technologies has transformed the way they work on deadline. Former *Newsday* photographer John Cornell, always an early adopter of technology for the paper, says that covering the annual visit of the tall ships to New York City around Independence Day changed completely in the span of twenty years: "The first time the big ships came to New York in the 1980s, I was dropping film off a rooftop in Brooklyn for a motorcycle driver to bring it back to *Newsday*. The last time . . . I had a laptop on top of my car and was sending images within minutes of the fireworks going up" (J. Cornell, personal communication, August 8, 2006).

The practical effect of such technology is to push back the deadline for getting photos to the newspaper. High-speed wireless connections give photographers as much as an extra hour to shoot and send pictures compared with working without them.

Cornell was one of the first newspaper photographers to experiment with wireless transmission of digital images in the 1990s. He remembers shooting digital photos of a murder scene in eastern Suffolk County, New York, miles from any office and not near any homes where he could borrow a phone line to send photos: "I sat in my car with the computer connected to the . . . phone and started the 25-minute transmission" (J. Cornell, personal communication, August 10, 2006). Cornell says the chief homicide detective was so fascinated by the process that he sat in the car and watched.

Cornell watched the technology evolve from year to year. In 1996, when TWA Flight 800 crashed off Long Island's South Shore, he and other photographers spent a month on top of a motel in Hampton Bays taking pictures of wreckage brought to the nearby Coast Guard station. Because wireless photo transmission wasn't available then, they used the phone lines in the rooms to transmit pictures.

One year later, when a small ceremony marking the anniversary of the crash took place on the beach, Cornell was there with a computer connected to a cell phone. "The most poignant moment was at dusk, [marking] when the plane went down. There were small lanterns on the beach with the survivors holding hands with the priest who was supporting them" (J. Cornell, personal communication, August 10, 2006). The beach was at least an hour away by car from the newsroom, so Cornell sent his pictures from the beach. "The only problem was the editor on that night did not realize it would take 15 minutes with that technology for each photo. We were right on the bitter edge of the deadline and all I remember is telling him, it is on the way" (J. Cornell, personal communication, August 10, 2006).

Three years later, in 2000, Cornell's wireless technology saved him again when he was in Florida covering the recount of paper ballots for the presidential election. More than twenty photographers on deadline fought for access to the five phone lines in the ballot-counting center to send images. Because of the need to share the lines, those photographers could send only a couple of images to their papers, Cornell says. "That left a group of photographers sitting in the parking lot with their laptops glowing and their cell phones connected sending more images back," he says.

When air cards became available for laptops, Cornell found himself transmitting photos while traveling at sixty miles per hour on the East Side of Manhattan in a presidential motorcade. "You no longer need to search for Wi-Fi hot spots or coffee shops," he says. The same technology allows sports photographers to send images back to the newsroom between innings during baseball games or during timeouts at other games. And photo editors can call photographers on their cell phones in the midst of a game to ask them to be on the lookout for certain shots.

The value of wireless communication devices for photographers was never more apparent than on September 11, 2001, in New York City, when the destruction of the World Trade Center brought the wireless network down with it as well.

> During 9/11 everything went down. We reached out to folks who had hard [land] line phones that worked. It was a frustrating day, mainly because we were so attached to our cell phones and quick communication and all of a sudden everything went down. It meant that most of us who could think on our feet just did what we thought we should do in the situation. Those who could not think ran around in circles. (J. Cornell, personal communication, August 10, 2006)

The *Times-Picayune* in New Orleans learned the same lesson when the city was flooded after Katrina and it had to abandon its building, publishing Pulitzer Prize–winning coverage while literally on the run. "One practical suggestion I would have is that communication systems are tremendously fragile," *Times-Picayune* editor Jim Amoss said in an interview with the Dart Center ("Covering the Recovery," 2006):

> We learned what it means to suddenly be bereft of everything that we use to communicate—cell phones, telephones, Internet hook-ups. In the aftermath of this, we've decided to acquire a variety of alternatives, from satellite phones to air cards, to cell phones and Internet connections and land lines. Because you just don't know what you're going to end up with. And as a newspaper you're so utterly dependant on it. Especially the Internet. Without the Internet you're lost as a newspaper. So you really have to figure out how you're going to continue if your normal means of connecting to the Internet is interrupted.

As Amoss found, even for editors, deskbound in the newsroom, wireless devices have changed the way they work. Because it is so much easier to find reporters by yanking on their electronic leashes, it is also easier to plan and coordinate coverage. No longer is it a mystery where reporters in the field might be or what they are doing; finding out is as easy as telephoning or emailing. Reporters in the field can communicate more easily with each other and their editor, ensuring that they don't duplicate each other's work if they are working together on a story.

Text communication is just as important as voice communication, according to Martin, now deputy marketplace editor at the *Wall Street Journal*. "The BlackBerry means we're able to be in constant contact with correspondents and editors around the world, which a truly global news operation couldn't really function without" (A. Martin, personal communication, August 7, 2006).

For early adopters and smart editors, wireless technology can also bestow a competitive advantage. In an article for Poynter Online in 2002, Peggy Carpenter-Johnson wrote how an air card–equipped laptop helped WIAT-TV of Birmingham to get a jump on the competition when former Ku Klux Klansman Tommy Blanton was retried in 2001 on charges of bombing the Sixteenth Street Baptist Church and killing four little girls in 1963. The judge refused to let anyone enter or leave the courtroom during testimony—and he banned cell phones from being near the courtroom. "Experience had taught us that getting a phone line during breaks would be next to impossible with so many reporters clamoring to call newsrooms during breaks" (Carpenter-Johnson, 2002). But with a laptop and an air card, WIAT reporter Stacey Norwood could send notes and stories to her newsroom without leaving her seat in the courtroom. That enabled her station to be first on the air with news from the trial for weeks—including the guilty verdict. "WIAT broadcast the verdict on the air before the doors of the courtroom could open for other reporters to get the word out" (Carpenter-Johnson, 2002).

Still, wireless technology is not yet perfect, and it has created some problems of its own. For example, because of the constant pressure to file first, file fast, and then update stories, there is even less time than before to worry about the nuance or fine details that make a story richer. "You just bang it out and live with the consequences. If it needs fixing, you fix it after the fact. . . . Print has become more like broadcast," according to one longtime wire service reporter (Anonymous, personal communication, August 9, 2006). The paradox, he says, is that devices designed to save time have, by their capabilities, given him less time to do his job.

And not all wireless devices are designed with the needs of journalists in mind. Phone companies, for example, give a higher priority to voice communication than to wireless data transmission. Some phone companies, in their infinite wisdom, will give the cell phone customer more space on the band and less space for the transmission of images or data. This can be hairy at large functions (J. Cornell, personal communication, August 10, 2006). During one fireworks show in New York City, Cornell said photographers in Queens had trouble using air cards to connect to send images, while those in New Jersey were fine.

But the advantages aren't hard to see.

Thirteen years and two months after the scene outside Joel Rifkin's mother's house, there was another spectacular murder arrest on Long Island. A Glen Cove man was charged with decapitating his neighbor, cutting her into pieces, and driving around for a while with the head in the trunk of his car. Once again, reporters flooded the neighborhood, talked to cops, and spread out to places where the suspect had lived and worked. But this time it was different.

The editors in the newsroom knew all the time where reporters were and what they were doing. We didn't have to wait to ask a reporter on the scene to talk with people at a different location or ask them questions about new information we had just learned. We called them on their cell phones and it was done. Compared with before, we were nimble and quick. Days later, when a reporter found people two hours away with vivid recollections of the suspect, he wasted no time looking for a payphone. He called from his car. And when another reporter chasing the suspect's past in California and Arizona found someone who feared for her safety after a run-in with him, there was no sense of panic when she called with the news late, right on deadline. There was no scramble to free a reporter for rewrite duty. The story came by email in less than twenty minutes.

References

Carpenter-Johnson, P. (2002, January 14). Adapt or die. Retrieved August 2006, from http://www.poynter.org/content/content_view.asp?id=4157.

Covering the recovery: An interview with New Orleans Times-Picayune editor Jim Amoss. (2006, May 4). Retrieved August 2006, from http://www.dartcenter.org/articles/personal_stories/amoss_jim.html.

Conclusion

ANYTIME, ANY PLACE

Mobile Information and Communication Technologies in the Culture of Efficiency

Sharon Kleinman

In the preface, I described a fairly routine example of anytime, any place connectivity: Alex's cell phone rang while she was cutting my hair, and she stopped to check the caller ID. After confirming that it wasn't someone calling from her daughter's school, she resumed cutting and confided that she turned her cell phone off when her daughter was with her. Carefully managing her anytime, any place connectivity had become second nature.

Users of mobile information and communication technologies (ICTs), as well as developers, nonusers, and even anti-users, continue to reconfigure where, when, and how these technologies are employed, as the authors of the chapters in this collection have masterfully illustrated. The diversity of the stories about mobile communication told in *Displacing Place* underscores that this exciting domain is literally a moving target for researchers.

The overarching theme of this book is that mobile ICTs are modifying geographic places, our relationships to place, and our relationships to one another in obvious and obscure ways. Cell phones, for example, are privatizing public space and altering our acoustical expectations. "There is a new kind of interaction of people who are wired displacing a public space," leading to uniquely twenty-first century regulatory, architectural, and planning challenges (Gumpert & Drucker, chapter 1, p. 19).

People are modifying the form, functions, symbolic meanings, and usage etiquette of mobile ICTs to meet changing needs, desires, and values. Mobile ICTs are modifying people, too. Interactions—mediated and face-to-face—affect us profoundly, down to our very neurons, as Yvonne Houy explained in chapter 4. As anytime, any place connectivity expands opportunities for when and where we work, play, learn, and relate to one another, our expectations and attitudes as well as how we go about the nitty-gritty aspects of living our lives and loving others are changed.

Tools for Life

While the life cycle effects that mobile ICTs—from production to disposal phases—are having on people, built environments, and the ecosystem are not yet fully understood, it is clear that in a very short time these technologies have become pivotal tools whose ubiquity and utility are distinguishing life in the early twenty-first century from earlier eras.

The cell phone is arguably becoming the Swiss Army knife of mobile ICTs (L. Crowley, personal communication, May 2, 2006; see also Levinson, 2004). This is an apt metaphor because in addition to their function for voice communication, cell phones are also handily used as text messengers, video and still cameras, calculators, day planners, address books, digital media players, personal navigation instruments, global position transmitters, memo pads, walkie-talkies, alarm clocks, video games, Internet connection devices for emailing and Web surfing, digital wallets, movie and television viewing machines, and more.

Above and beyond their multifunctionality, cell phones, as well as other mobile ICTs, are increasingly salient elements in the "bricolage" of life, the mix-matching and overlaying of objects that produce cultural identities (du Gay et al., 1997). People handle, use, accessorize, and wear these tools in innovative and symbolic ways that visibly and audibly communicate information about themselves and their lifestyles. The Swiss Army knife metaphor is applicable here as well, as there is a long tradition of workers and outdoorsmen wearing their knives in cases on their belts.

As more and more people worldwide adopt mobile ICTs, and connectivity becomes increasingly seamless through interoperability standardization that enables devices to "talk" with one another, our opportunities and capabilities for multitasking in the hybrid space Yvonne Houy calls the "Metro/Electro Polis" in chapter 4 are expanding.

Multitasking in physical and virtual spaces has been possible since the 1800s, with the advent of the telegraph and then the telephone (Standage, 1998). By the 1950s, early car phones enabled a new form of multitasking in two spaces involving mobile communication. A physician and a bank-on-wheels in Stockholm were among the first to adopt the eighty-pound car phones that reportedly drained the car's battery after two calls (Lacohee, Wakeford, & Pearson, 2003). By the mid-1960s, with the advent of transistors, mobile phones shrunk to briefcase size, but they were still heavy and required cars to transport them. By the 1990s, technological advances leading to miniaturization and longer battery life enabled developers to further transform car phones, which had been used primarily by professionals, into lighter handheld devices, which were then marketed to a wide audience for everyday uses (Agar, 2003). When we consider today's cell phones' magnificent panoply of image, sound, text, and memory capabilities—which Richard Olsen highlighted in chapter 9—they seem light-years ahead of the telegraph and telephone of the 1800s, the car phones of the 1950s, and even the cell phones of the 1990s!

Today's mobile ICTs enable colleagues to virtually enter each other's workspaces to communicate and collaborate easily and efficiently, facilitating the just-in-time knowledge sharing, problem solving, and decision making that have become hallmark organizational practices in the twenty-first century, as Calvert Jones and Patricia Wallace explained in chapter 10. Anytime, any place access to colleagues, data, and documents alters not only how people work, but also where and when; boundaries are blurred between public and private domains as well as between work and leisure time, which can increase workers' stress levels (Chesley, 2005). Many people feel compelled to stay connected and accessible via their mobile ICTs all the time and everywhere because this equipment supports anytime, any place connectivity. Thus, an implicit drawback to having an employer-owned cell phone for some people is that although they are accessible to their family and friends when they are working, they are also accessible to colleagues when they are not working, in theory if not always in practice (Ling & Haddon, 2003). A third of all U.S. employees are in contact with work during leisure time once a week or more (Galinsky et al., 2005). Think of the now familiar scenario of a group of people enjoying a meal together at a restaurant when someone's cell phone rings. The cell phone owner quickly checks the caller ID and announces to the group, "I have to take this call—it's my boss/colleague/client."

Beyond their work applications, mobile ICTs have become central for nurturing and sustaining relationships, especially over great distances. This is dramatically illustrated when members of the U.S. military serving in remote areas video teleconference with their loved ones in the United States (Wielawski, 2005). In one poignant case, a soldier in Iraq talked with his wife six thousand miles away in West Virginia during her labor and witnessed the joyous occasion of their daughter's birth via a live Internet video and satellite audio link ("Internet Takes Soldier into Delivery Room," 2005). Then again, because of mobile technologies, many people are witnessing traumatic events, and this has important psychological implications as well (E. Perlswig, M.D., psychiatrist, personal communication, November 1, 2006).

During tragedies in the early twenty-first century, mobile ICTs have become vital lifelines: emergency-response personnel have drawn on them to coordinate operations; friends, family, and coworkers have used them to contact one another; and professional journalists, as well as nonspecialist citizen journalists, have employed them to visually and orally communicate breaking news and eyewitness accounts, enabling people all over the world to experience catastrophes in other places in a surreal way. Perhaps anytime, any place connectivity promotes empathy under these circumstances.

Social Networking and the Sociability-Productivity Paradox

People are increasingly using mobile ICTs in public places, such as parks, hotels, libraries, bars, stores, shopping malls, restaurants, cafés, museums, and sports arenas, as wireless broadband Internet access becomes progressively more available and additional devices are able to tap into it. The number of public wireless access points (hot spots) worldwide exceeded 100,000 in January 2006; of these about 8,000 offer free access (Hamblen, 2006). Cities and towns throughout the United States are enthusiastically pursuing wireless broadband Internet initiatives, as Harvey Jassem detailed in chapter 2.

When people no longer tethered to large, heavy machines in their offices or homes do their work in public places using lightweight, portable equipment, they increase their opportunities for interacting face-to-face with others. Impromptu social interactions can promote a sense of well-being. I have experienced this myself as a professor: I sometimes write and grade papers at Koffee, a quintessentially twenty-first-century café in my hometown that provides its customers with free wireless broadband Internet access and surge-protecting power strips for their mobile gear. Many patrons at Koffee work on laptops and have cell phones and digital media players, most noticeably the iconic iPod, poised next to their computers. At this café, I feel part of the place-based community even when I am working at a table alone (compare Oldenburg, 1999, 2001). When I run into friends I know from other contexts or friendly acquaintances I have seen there before, it is an added bonus.

A friend who also enjoys working at Koffee sums up what I have come to think of as a sociability-productivity paradox: mobile ICTs have greatly expanded opportunities for where and when people work, which can increase productivity, yet they have also expanded opportunities for face-to-face and virtual social interactions, which can reduce productivity (K. Cooke, personal communication, March 12, 2006). When "information-haves" become regulars at places like Koffee and use their mobile ICTs for work there, they become especially susceptible to reducing their productivity over time because, as they get to know other regulars, the frequency and depth of their face-to-face interactions tend to increase. Yet these impromptu face-to-face interactions can lead to unanticipated positive "consequences" as well, ranging from new information to new friendships.

Conversely, some mobile ICT users don't seem to even notice others or the happenings around them in public, as they are profoundly disengaged from their place-based experiences while talking on cell phones; listening to music or watching movies on digital media players; or working, emailing, Web surfing, gaming, viewing pornography, or gambling on laptop computers or personal digital assistants (PDAs). We have to wonder about the long-term ramifications.

Of course, concerns about mobile ICTs having negative psychological and social ramifications are not new. In the 1980s, misgivings were expressed regarding the Sony Walkman, which was launched internationally in 1979 (du Gay et al., 1997). Some argued that the Walkman was distracting, atomizing, and alienating, and that it was eroding public life because Walkman users absorbed in their private listening experiences were tuning out their place-based experiences. Countering these notions, users argued that music provided an enhancing soundtrack for their place-based experiences and that listening to books, foreign language lessons, and study guides on tape was a way for Walkman users to tap into culture rather than to escape it (du Gay et al., 1997).

However, since the 1980s, there have been tremendous quantitative and qualitative changes in mobile communication. In the twenty-first century, mobile ICTs with an astounding array of capabilities have become ubiquitous. What's more, it is becoming increasingly common for people in public places to prioritize their mobile ICT activities over their place-based experiences. "Indeed, these days even in the public street a stranger seeking directions in a new city may have to wait patiently for a passerby without a cell phone, headset, or handheld device or risk interrupting another's virtual reality, the new social faux pas" (Bugeja, 2005, p. 2). When I described this scenario to a friend, she wryly suggested that if people seeking directions had cell phones or GPS-enabled PDAs, they could call for directions or look them up rather than interrupt others on the street (K. Cooke, personal communication, March 13, 2006).

Split engagement in physical and virtual spaces can lead to dire consequences for mobile ICT users as well as bystanders. Car drivers are four times more likely to be injured in a crash when they are talking on a cell phone, even if they are using a hands-free device ("Put Calls on Hold," 2006). Talking on a cell phone while driving decreases drivers' reaction times and increases the risk of crashing to a similar level as driving with a blood-alcohol level that exceeds the legal limit (Seo & Torabi, 2004). Remarkably, during any daylight moment, 10 percent of all drivers on U.S. roads are using cell phones, according to the National Highway Transportation Safety Administration (Glassbrenner, 2005).

Unplugging

Around 100 B.C., the Roman philosopher Publilius Syrus warned that "to do two things at once is to do neither." In the twentieth century, Mahatma Gandhi remarked that "there is more to life than increasing its speed." Yet multitasking with mobile ICTs has become a hallmark practice of the twenty-first century, and the pace of everyday life is being sped up as more people put these technologies to work:

> Succeeding in today's economy requires lightning fast reflexes and the ability to communicate and collaborate across the globe. . . . Globalization and the Internet create great new opportunities, but they also ratchet up the intensity of competition and generate more work . . . 25% of executives at large corporations say their communications—voicemail, e-mail, and meetings—are nearly or completely unmanageable. (Mandel et al., 2005, p. 60)

While there are many instances when carefully considered communication or actions are warranted, it is easy to lose sight of this and communicate or act without delay because mobile ICTs are so handy—literally in our hands (Nyíri, 2003). A veteran public relations professional told me that she knew it was time to leave her high-powered corporate job and pursue something new—teaching part time at a university—when she found herself repeatedly checking her BlackBerry at the movies, much to her friends' and family's dismay (B. Levy, personal communication, January 2006). These machines are sometimes jokingly referred to as "CrackBerries" because users become notoriously addicted to them. This woman is part of a growing segment of society self-regulating their connectivity to achieve more balanced lifestyles.

While many people are compelled to maintain anytime, any place connectivity for professional reasons (physicians on call, for instance, as Keith J. Ruskin explained in chapter 11) or for personal reasons (parents of young children, for example), others have more latitude regarding their connectivity. However, in our culture of efficiency, many have grown accustomed to—and in some cases addicted to—multitasking with mobile ICTs. Awareness of the rich technical capabilities of the assorted equipment adorning clothing and carried in pockets, purses, and briefcases is incentive enough for some people to use this gear, no matter the time or place. When people make work calls while they are driving or send email and text messages while they are eating lunch, the tempo of life gets ratcheted up, and something is lost.

Before *Displacing Place* went to press, I shared Andrew Smith's chapter about how mobile ICTs are transforming the practice of journalism with Robert Miller, a photographer who works for the *New York Post*. I emailed it to him from my laptop computer, of course, in a matter of seconds, using the Wi-Fi connection at a local coffee shop. After reading the chapter, Miller told me that editors sometimes expect photos from him "in 'real time' because they are watching the same event on *live* television in the office. To meet tight deadlines, by taking the time to transmit photos back from the field while [an event] is unfolding, I may actually miss what is happening while . . . editing and sending photos rather than shooting [the event] in its entirety" (personal communication, October 28, 2006).

Miller concluded that there are drawbacks to mobile communication, as is the case with most things, but that he would never want to go back to the way it was before mobile ICTs were widely employed in journalism, echoing Smith and the journalists and photographers quoted in chapter 14. "I used to have to travel back to the office to develop film. For some assignments, I had to FedEx film overseas, which took days and left the editing up to someone else. Now, my car is my darkroom, and I can wirelessly via a laptop computer and air card send a picture anywhere most times in under a minute, and as a result I am the one who does the initial edit. I choose which pictures to send back to the office from my entire shoot" (R. Miller, personal communication, October 28, 2006).

As mobile ICTs keep evolving and proliferating, they will become even more integrally woven into our lives than they are already, and we will multitask in physical and virtual spaces with greater frequency and ease. For people with physical limitations that affect their sight, hearing, speech, writing, or mobility, and for those with certain psychological problems, mobile ICTs will continue to be lifelines, literally, that make it possible for them to communicate with others and participate more fully in the world. For many people, the usefulness and convenience of these technologies and the lifestyle flexibility that they support will continue to outweigh potential or realized negative consequences. Yet the effects are still unfolding.

We have to continue exploring questions raised in *Displacing Place*:

- How are mobile ICTs altering the way people work, play, learn, and relate to one another?
- How are they impacting interpersonal relationships, communities, and the environment?
- How are they changing urban and rural life?
- How are they contributing to democratic discourse, social activism, and awareness and understanding of current events?
- How can they be employed to narrow the knowledge and socioeconomic status divides?
- How are people using, modifying, and resisting them to meet changing needs, desires, and values?
- How can we promote environmental and human health in all places as we continue to imaginatively devise and adopt technologies for displacing place?

In the years to come, we must energetically pursue opportunities for using mobile information and communication technologies in ways that improve everyday life and enhance our connections to one another.

References

Agar, J. (2003). *Constant touch: A global history of the mobile phone*. Cambridge: Icon.

Bugeja, M. (2005). *Interpersonal divide: The search for community in a technological age*. New York: Oxford University Press.

Chesley, N. (2005). Blurring boundaries? Linking technology use, spillover, individual distress, and family satisfaction. *Journal of Marriage and Family*, 67, 1237–1248.

du Gay, P., Hall, S., Janes, L., Mackay, H., & Negus, K. (1997). *Doing cultural studies: The story of the Sony Walkman*. London: Sage.

Galinksy, E., Bond, J. T., Kim, S. S., Backon, L., Brownfield, E., & Sakai, K. (2005). Overwork in America: When the way we work becomes too much. New York: Families and Work Institute. Retrieved September 18, 2006, from http://familiesandwork.org/.

Glassbrenner, D. (2005). *Driver cell phone use in 2005: Overall results*. (DOT HS 809 967). Washington, D.C.: National Highway Transportation Safety Administration.

Hamblen, M. (2006, February 8). Boston considering public Wi-Fi. Retrieved March 1, 2006, from http://www.computerworld.com/printthis/2006/0,4814,108479,00.html.

Internet takes soldier into delivery room. (2005, August 9). Retrieved September 18, 2006, from http://www.culturecraze.ca/print-1461.html.

Lacohee, H., Wakeford, N., & Pearson, I. (2003). A social history of the mobile telephone with a view of its future. *BT Technology Journal*, 2(3), 202–211.

Levinson, P. (2004). *Cellphone: The story of the world's most mobile medium and how it has transformed everything!* New York: Palgrave Macmillan.

Ling, R., & Haddon, L. (2003). Mobile telephony, mobility, and coordination of everyday life. In J. E. Katz (Ed.), *Machines that become us* (pp. 245–265). New Brunswick, NJ: Transaction Publishers.

Mandel, M., Hamm, S., Matlack, C., & Farrell, C. (2005, October 3). The real reasons you're working so hard. *Business Week*, 3953, 60.

Nyíri, K. (2003). Introduction: From the information society to knowledge communities. In K. Nyíri (Ed.), *Mobile communication: Essays on cognition and community* (pp. 11–23). Vienna: Passagen Verlag.

Oldenburg, R. (1999). *The great good place: Cafes, coffee shops, bookstores, bars, hair salons and other hangouts at the heart of a community*. New York: Marlowe and Company.

———. (2001). *Celebrating the third place: Inspiring stories about the "great good places" at the heart of our communities*. New York: Marlowe and Company.

Put calls on hold to prevent crashes. (2006). *Liberty Lines*, 10(1), 5.

Seo, D. C., & Torabi, M. R. (2004). The impact of in-vehicle cell-phone use on accidents or near-accidents among college students. *Journal of American College Health*, 53(3), 101–107.

Standage, T. (1998). *The Victorian Internet: The remarkable story of the telegraph and the nineteenth century's on-line pioneers*. New York: Walker & Company.

Wielawski, I. M. (2005, March 15). For troops, home can be too close. *New York Times Magazine*, F1, F6.

ABOUT THE CONTRIBUTORS

BRYAN BALDWIN is a Ph.D. student in the Department of Communication at the University of Massachusetts at Amherst, assistant to the president for communications at Bridgewater State College, and formerly an economic officer with Foreign Affairs and International Trade Canada. He received his B.A. in political science and economics from the University of Massachusetts at Amherst and his M.A. in political communications from Emerson College.

GENE BURD is associate professor of journalism at the University of Texas at Austin. He teaches news reporting and writing and does research on cities and communication. He was a journalist at the *Kansas City Star, Houston Chronicle*, and *Albuquerque Journal*, and at suburban newspapers in Los Angeles and Chicago; taught at the University of Minnesota and Marquette; attended UCLA, Iowa, and Northwestern (Ph.D., 1964); was one of the last residents of Jane Addams's Hull-House in Chicago; and is the founding benefactor of the Urban Communication Foundation.

SUSAN J. DRUCKER is professor in the Department of Journalism/Media Studies in the School of Communication at Hofstra University. She is an attorney, editor of the *Free Speech Yearbook* and *Qualitative Research Reports in Communication*, and series editor of the Communication and Law series for

Hampton Press. She is the author and editor of six books and over eighty-five articles and book chapters, including *Voices in the Street: Explorations in Gender, Media, and Public Space* and two editions of *Real Law @ Virtual Space: The Regulation of Cyberspace* (1999, 2005) with Gary Gumpert. She is a recipient of the Franklyn S. Haiman Award for distinguished scholarship in freedom of expression. Her work examines the relationship between media technology and human factors, particularly as viewed from a legal perspective. She is treasurer of the Urban Communication Foundation, a not-for-profit organization supporting research on communication and the urban condition.

GARY GUMPERT is emeritus professor of communication at Queens College of the City University of New York and co-founder of Communication Landscapers, a consulting firm. His publications include *Talking Tombstones and Other Tales of the Media Age*; three edited volumes of *Inter/Media: Interpersonal Communication in a Media World* published by Oxford University Press; *Voices in the Street: Explorations in Gender, Media, and Public Space*; and *The Huddled Masses: Immigration and Communication*. He is a frequent contributor to the International Institute of Communication publication *InterMedia*. His most recent book is *Real Law @ Virtual Space: The Regulation of Cyberspace* (2nd edition). He is a recipient of the Franklyn S. Haiman Award for distinguished scholarship in freedom of expression. He is president of the Urban Communication Foundation, a not-for-profit organization supporting research on communication and the urban condition, and president of the U.S. chapter of the International Institute of Communication. His primary research focuses on the nexus of communication technology and social relationships, particularly looking at urban and suburban development, the alteration of public space, and the changing nature of community.

JARICE HANSON is professor in the Department of Communication at the University of Massachusetts at Amherst and currently holds the Verizon Chair in Telecommunication as visiting professor in the School of Communications and Theater at Temple University. Her research focuses on the social impact of digital technologies. She has published fifteen books and numerous articles on media, society, and cultural interpretations of mediated interaction.

YVONNE HOUY has eclectic research interests. She earned her Ph.D. from Cornell University in 2002 by writing an interdisciplinary dissertation on advertising, politics, and film from the 1920s through the Second World War, while her most recent publication examines how urban spaces are being imaginatively and meaningfully transformed by mobile and Internet technologies. She was webmistress for the academic organization Women in German for several years, and until recently penned a regular column on online resources and Internet issues. Currently, she teaches cultural studies courses at the Honors College of the University of Nevada at Las Vegas and informally researches how her toddler reacts to various communication devices while maintaining attachments to far-flung grandparents.

HARVEY JASSEM has written on telecommunication law and policy for over thirty years. He specializes in analyses of emerging media and the interplay of regulation and the shaping of those new media. He is executive board member of the Urban Communication Foundation and of the Fellows program of the American Council on Education, has served on multiple civic and governmental commissions, and was founding director of Loyola University Chicago's School of Communication, Technology and Public Service, and the University of Hartford's School of Communication, where he is presently a member of the faculty. He received his Ph.D. in communication in 1977 from the University of Wisconsin at Madison.

CALVERT JONES is a graduate student in international relations at the University of Cambridge. She holds a master's degree from the School of Information at the University of California at Berkeley. Her research focuses on the evolution of nonstate actors and their technology-enabled networked power. She has conducted research for the Markle Foundation on American intelligence reorganization and with faculty at Stanford University's Hoover Institution on international cooperation against transnational threats. Her professional experiences include NGO assignments to Vietnam and the Balkans, where she worked with a transnational network in Kosovo, Bosnia-Herzegovina, and Croatia.

JULIAN KILKER is associate professor of emerging technologies at the Greenspun School of Journalism and Media Studies at the University of Nevada at Las Vegas. He received a B.A. in physics from Reed College and an M.S. and Ph.D. in communication from Cornell University. His research focuses on the intersection of social interaction, technology, and design, particularly in relation to communication resources, with current projects examining user control and unintended consequences in emerging media and comparative life cycles in traditional and emerging media. His work has been published in numerous journals, including *Management Communication Quarterly*, *Convergence*, *Iterations*, and *IEEE Technology and Society*.

SHARON KLEINMAN is professor of communications at Quinnipiac University. Her research focuses on the history and social implications of communication technologies and on issues concerning online and place-based communities. She has received numerous academic awards, including the Outstanding Faculty Scholar Award from Quinnipiac University and the Anson Rowe Prize from Cornell University. She holds a B.A. in English and American literature from Brandeis University and an M.S. and Ph.D. in communication from Cornell University. An avid mountain biker, photographer, and yoga practitioner, she lives in New Haven, Connecticut.

PENNY A. LEISRING is associate professor of psychology at Quinnipiac University. She received her Ph.D. in clinical psychology from the State University of New York at Stony Brook in 1999. Her clinical interests focus on the prevention and reduction of aggressive behavior in adults and children. She conducts research examining male- and female-perpetrated relationship aggression, parental discipline styles, and child behavior problems.

JULIE NEWMAN is director of the Office of Sustainability at Yale University. She also worked in the field of sustainability and higher education at Tufts University and the University of New Hampshire. In 2004, she co-founded the Northeast Campus Sustainability Consortium, which she now chairs. This consortium was established to advance education and action for sustainable development on university campuses in the northeastern United States and Canadian Maritime region. Her research has focused on the role of decision-making processes and organizational behavior in institutionalizing sustainability into higher education. She holds an M.S. in environmental policy from Tufts University and a Ph.D. in natural resources and environmental studies from the University of New Hampshire.

RICHARD OLSEN is a faculty member at the University of North Carolina at Wilmington. He teaches in the areas of rhetorical theory, popular culture, and research methods. His research centers on how various trends and artifacts of popular culture shape and reflect cultural values. He has published work on the popularity of sport utility vehicles (SUVs), the NBA draft, and MTV.

GARY PANDOLFI is instructional technologist and adjunct associate professor of English at Quinnipiac University, where he supports faculty in the School of Communications and College of Liberal Arts as they use technology for teaching and learning. He has taught English to students from middle school to university for over thirty years. As an instructional designer at McGraw-Hill, he designed educational software to complement textbooks. He has written online courseware for SkillSoft Corporation as an instructional designer at Business Performance Technology. He received his B.A. from Hamilton College and M.A.L.S. from Wesleyan University.

KEITH J. RUSKIN is professor of anesthesiology and neurosurgery at Yale University School of Medicine. His research interests include medical informatics and clinical applications of communication technology. He has published numerous articles and book chapters on information technology. He holds a commercial pilot certificate with airplane single- and multi-engine ratings.

ANDREW SMITH is deputy Long Island editor for *Newsday*, the largest suburban daily newspaper in the United States. Before that, he was a reporter for twenty years at several newspapers. He has covered courts, transportation, several troubled nuclear power plants, a nuclear weapons laboratory, politics, and government. He and Earl Lane won the White House Correspondents Association award for national reporting for their series on the difficulty of nuclear waste disposal. He was part of *Newsday*'s coverage of the Flight 800 disaster that won the Pulitzer Prize. He also once wrote a story on deadline that rhymed, including the quotes.

PATRICIA WALLACE is senior director of information technology and distance education at the Johns Hopkins University Center for Talented Youth. Her research and writing focus on the role of advanced technologies in society and human behavior. Her recent books include *The Internet in the Workplace: How New Technology Is Transforming Work* and *The Psychology of the Internet*, both published by Cambridge University Press. She received her Ph.D. in psychology from the University of Texas at Austin and also holds an M.S. in computer systems management.

MATTHEW WILLIAMS is lecturer in the School of Social Sciences at Cardiff University and the independent academic advisor on e-crime to the Welsh Assembly Government. He has conducted research and published extensively in the areas of cyber-crime; online and digital research methodologies and sexuality; and policing and criminal justice. He was on the board of directors for the Association of Internet Researchers (AoIR) and is on the editorial board for *Sociological Research Online* and the *Internet Journal of Criminology*, and is book and associate editor for *Criminology and Criminal Justice*. He is the author of *Virtually Criminal: Crime, Deviance and Regulation Online* (Routledge, 2006) and numerous journal articles. His recent research includes "Ethnography for the Digital Age" (Economic and Social Research Council [ESRC] in the UK), "Methodological Issues for Qualitative Data Sharing and Archiving" (ESRC), and "Counted Out 2: A Survey of Lesbian, Gay and Bisexual Communities in Wales" (Welsh Development Agency and Stonewall Cymru).

INDEX

9/11 terrorist attacks, 221
21st Century Classrooms Act, 82

absent presence, 50
acoustical space, 13–14
addiction to MCTs, 50, 230–31
adoption of new technology, 141
advertising
  amount spent on by mobile phone companies, 142
  appeal to needs in, 145–53
  belongingness in, 147–48
  "choice" in, 149–50
  and creation of meanings, 142
  creation of perpetual anxiety and discontent in, 152
  and fantasy, 143–44, 151
  fantasy theme analysis, 146
  focus on ideals, 144
  goals of, 142–43
  minor themes in mobile phone ads, 144–45
  "more" in, 150–51
  nature theme in, 145
  rhetorical vision in, 144, 146, 150–53
  safety and security theme in, 146–47
  self-expression theme in, 148–50
  transcendence theme in, 150–53
  understanding and use of things, 142
  utopian appeal of mobile phone ads, 151
affect regulation, 67–68
Ainsworth, Mary, 63–64
air cards, 218–19, 221, 222
a-location, 12
Al Qaeda, use of MCTs, 170
Altheide, David L., 152
Amazon.com, 111
America Online, 34
Amoss, Jim, 221
A&M Records, Inc. v. Napster, 98
Amusing Ourselves to Death (Postman), 154
Anderson, Christopher, 190
Andersson, Gerhard, 192
anxiety disorders and e-therapy, 193–94
"anytime, any place" connectivity
  disadvantages of, 231
  in networked organizations, 227

Apple Computers, 85
architecture, virtual, 45
Archos Recorder, 115
ARPANET, 108–9
attachment
  affect regulation, 67–68
  lack of emphasis on in CMC research, 62
  and MCTs, 60–62
  physical co-regulation, 67–68
  working models of, 64, 66–67
attachment figures, 66–67, 68
attachment styles, 63–64, 66
attachment theory
  and neurobiology, 65–67
  research on, 63–64
attunement, 70, 72
auditory aggression, 13–14
avoidant attachment style and use of MCTs, 68

backchannels, 166–68
backstage negotiation, 166–68
banks, cost of phishing to, 98
Basal Convention, 79–80
basic needs in mobile phone ads, 145–53
Baym, Nancy, 61, 62–63
belongingness as theme in mobile phone ads, 147–48
Berman Bill, 99, 102
Bird, Frederick L., 28
Birmingham, Alabama, 222
bodies, 50–51
Boing Boing, 114
Boisvert, Jean-Marie, 192
Boone, Douglas, 29–30
bootstrapping, 109
boredom, 153
Bormann, Earnest, 144
Boston, 32–33
bot-nets, 97
Bouchard, Stéphane, 193
Bowlby, John, 63
brain, 65–66
broadband, 22–23
Brummett, Barry, 143, 151, 152
Bucy, Erik P., 136
bullying, use of mobile phones in, 101
Burke, Kenneth, 143

Caine, Chris, 30
California, e-waste disposal program in, 81–82
Canadian Broadcasting Act of 1991, 130
Canadian Broadcasting Corporation (CBC), 129, 130
Capner, Maxine, 198
Carnegie Commission, 129
Carpenter-Johnson, Peggy, 222
CBC (Canadian Broadcasting Corporation), 129, 130
Celio, Angela, 194
cell phones. *see* mobile phones
chat groups, e-therapy via, 191
"choice" in advertising, 149–50
cities
  as communication networks, 44
  decline of walled cities, 41
  and displacement, 44
  historical concern over demise of, 39
  impact of mobile communication on, 40–41, 43
  impact of muni Wi-Fi on, 35
  impact of new media on, 43
  modern design of, 51
  and physical sensations in space, 51
  reasons for providing muni Wi-Fi, 24–25
  space and time in, 50
  survival of, 39–53
  as switchboard, 8
  tactile reality, 51
  zoning in, 16–18
  . *see also* community
Clarke, Richard, 95
classrooms, technology in, 208–14
clinical supervision, online, 200–201
cloning, 99
CMC (computer-mediated communication)
  and online social life, 60–61
  research on, 62–63
CNET, 111–12

collaboration, 116, 118
collective action and MCTs, 163–65
communication, emotional, 69–71, 72
. *see also* nonverbal communication
communication, mobile
social implications of, 11–18
updated paradigm of, 9
. *see also* ICTs; MCTs
communication, nonverbal. *see* nonverbal communication
Communications Decency Act, 102
communication standards, need for in healthcare, 182–83
communication technology as nonplace event, 8
community
and disconnection of person from place, 8
extension of by mobile communication, 44–45
and geographical place, 46
impact of muni Wi-Fi on, 35
on the Internet, 46–47
in mobile phone ads, 146, 147
mobile phone as way to cut users off from, 153, 154
and participation in public life, 14
and podcasts, 138
transformation to nonplace orientation, 8
. *see also* cities
computer games, modification of, 114–15
computer-mediated communication (CMC). *see* CMC
computers, laptop, 9
Concurrent Versioning System (CVS), 116
Connor-Greene, Patricia, 190–91, 198
control, 106, 111–12
. *see also* meta-control
Convention on Cybercrime (2001), 102
Coolidge, Calvin, 28
Cornell, John, 219–21, 223
Corporation for Public Broadcasting (CPB), 129
course management systems. *see* instructional technology
CPB (Corporation for Public Broadcasting), 129
Creative, 111
cryptoviral extortion, 97
Curry, Adam, 127
CVS (Concurrent Versioning System), 116
The Cybercities Reader (Graham), 13
cyber-crime
classification of, 93–94
cyber-espionage, 95
cyber-fraud, 98–99
cyber-terrorism, 95, 184–85
cyber-theft, 98–99
cyber-violence, 99–101
hacking, 94–96
jurisdictional inconsistencies, 93
legal reactions to, 93
malware, 96–97
regulating, 102
temporal and spatial dimensions of, 92–93
cyber-espionage, 95
cyber-fraud, 98–99
cyber-stalking, 100–101
cyber-terrorism, 95, 184–85
cyber-theft, 98–99
cyber-violence, 99–101

Daughton, Suzanne, 144
The Death and Life of Great American Cities (Jacobs), 17
DeCSS, 113
de Haan, Hein, 192
de Jong, Cor, 192
Dell Computers, 85
Demiris, George, 196
denial-of-service attacks, 97
developing nations, importance of MCTs in, 165
deviance. *see* cyber-crime
Diffusion of Innovations (Rogers), 141
digital images, wireless transmission of, 220
digital media players
and control, 111–12
influence of, 106
modification of, 115–18
Digital Millennium Copyright Act (DMCA), 98–99

disaster recovery and MCTs, 164–65
disconnection
anxiety caused by, 16
from space, 50–51
discussion boards, e-therapy via, 191
disengagement from place-based experience, 229
displacement, 11–12, 44
DMCA (Digital Millennium Copyright Act), 98–99
downloading, mobile phone as device for, 152, 153
downloads as part of belongingness, 148
driving and using mobile phones, 230
DVDs
copying of, 113
restrictions on viewer control, 111

early adopters, influence of, 113–14
Earthlink, 31–32
eating disorders and e-therapy, 194–95
economic development and muni Wi-Fi, 25
education. *see* instructional technology
electricity, 27–29
electronic health records (EHRs), 186
electronic waste. *see* e-waste
Electronic Waste Recycling Promotion and Consumer Protection Act, 82–83
Electropolis, 60
email, therapy via, 190–91
emotional communication, 69–71, 72
. *see also* nonverbal communication
emotional connection, importance of, 70–71
Engelbart, Douglas, 109
environmental hazards of e-waste, 78, 79
environments, contraction of, 12
EPR (Extended Producer Responsibility) regulations, 84
e-therapy
advantages of, 195–97
and anxiety disorders, 193–94
clinician time, 198–99
confidentiality issues, 197
convenience of, 195
cost of, 195–96
disadvantages of, 197–99
and eating disorders, 194–95
ethical issues, 197, 199
increase in use of, 189
lack of physical presence, 198
legal issues, 197–98, 199–200
licensure, 197–98
and managed care companies, 196
Online Anxiety Prevention Program, 193
online clinical supervision, 200–201
populations using, 191–92
and relapse prevention, 191
research on, 192–95
Student Bodies, 194–95
via chat groups, 191
via discussion boards, 191
via email, 190–91
European Union
e-waste regulations, 83–84
regulation of Internet, 102
European Union Action Plan on Promoting the Safe Use of the Internet, 102
e-waste
corporate policies and procedures, 85–86
hazards of, 78, 79
increase in, 78
international trafficking of, 79–80
landfill needed for, 79
producers' responsibility for disposal of, 83–84
recycling of, 78–79, 80, 83
regulations in European Union, 83–84
regulations in United States, 81–83
responsibility for, 86–87
take-back programs, 85
universal waste regulations, 81
exclusiveness as theme in mobile phone ads, 148
experience
and human brain, 65–66
and internalized working models of attachment, 64, 66–67
experiential insight, 70
expertise and networked organization, 165–66
Extended Producer Responsibility (EPR) regulations, 84

Extensible Markup Language (XML), 183
face-to-face communication, 61–62
family as theme in mobile phone ads, 147
fantasy theme analysis, 143–44, 146
fantasy themes in mobile phone ads, 146–48
fashion, technology as, 47, 148
Federal Electronics Challenge, 86
Francisco, M. J., 27
fraud, financial, 98–99
friendship as theme in mobile phone ads, 147

Gallese, Vittorio, 70
Gammon, Deede, 200
Gandhi, Mahatma, 230
geography, importance of, 46
global positioning system (GPS), 10
Glueckauf, Robert, 192, 196
Goffman, Erving, 107, 166
Google, 32
Gordon, Barry, 198
government
  provision of Wi-Fi (*see* muni Wi-Fi)
  role as utility provider, 29
GPS (Global Positioning System), 10
Graham, Stephen, 13
Green, Nicola, 147
Gregson, Kimberly S., 136
grief, pathological, 193–94

hacking
  cyber-espionage, 95
  cyber-terrorism, 95, 184–85
  targeting of mobile phones, 96
  types of, 94–95
  use of ICTs, 95–96
hacktivists, 94
Haddon, Leslie, 142
Haefner, James, 142
Hammerslet, Ben, 127
Hargrove, David, 200
Harper, Marion, Jr., 145
Hart, Roderick, 144
hazardous materials in e-waste, 79
health and physical proximity, 59, 67
healthcare
  availability of medical information, 186
  communication technology used in, 177–80
  cyber-terrorism and, 184–85
  electronic health records, 186
  information, types of, 176–77
  Massachusetts E-Health Collaborative, 186
  medical equipment, MCTs' interference with, 180–82
  National Health Information Network (NHIN), 186
  need for communication standards, 182–83
  pagers, 177–78
  patient safety, 185
  radiology, use of telementoring in, 177
  remote ICU monitoring, 177
  smart cards, 184
  telementoring, 177
  telemetry, 176
  therapy (*see* e-therapy)
  use of mobile phones, 178–79
  VoIP in, 179–80
  WLANs in, 179
Health Insurance Portability and Accountability Act of 1996 (HIPAA), 183–84
Health Level Seven (HL7), 183
"Herbless," 94
hierarchical organizations, 160
HIPAA (Health Insurance Portability and Accountability Act), 183–84
HL7 (Health Level Seven), 183
Hoover, Herbert, 28
Hopps, Sandra, 192
hospitals. *see* healthcare
Hou Tng, Tai, 148
Houy, Yvonne, 226
humans, mobility of, 7–8
Hurricane Katrina, 218–19, 221

IAER (International Association of Electronics Recyclers), 79, 86
IBM's support for muni Wi-Fi, 30
ICTs (information and communication technologies)

complacency regarding security of, 97
and cultural identity, 226
and cyber-fraud, 99
disposal of, 77, 78 (*see also* e-waste)
early use of, 108–9
flexibility of, 105
and hacking, 95–96
malware attacks on, 97
misuse of, 93 (*see also* cyber-crime)
mobile (*see* MCTs)
mobility of, 61–62
modification of, 109 (*see also* modding; modding communities)
and therapy (*see* e-therapy)
use of in cyber-violence, 101
. *see also* MCTs; POD devices
ICU, remote monitoring, 177
ideals, advertising's focus on, 144
identity and place, 48
images, still, 71
independence in mobile phone ads, 146
infinite, fantasy of, 151
information, access to, 24
infrastructure, government's role in providing/regulating, 27–29
instructional technology
advantages/disadvantages to students, 212–14
importance of understanding technology used, 210, 211
keeping focus on course objectives, 209–10
maximizing classroom time, 211
intellectual property, theft of, 98–99
Interapy, 193
International Association of Electronics Recyclers (IAER), 79, 86
Internet
access, demand for, 23
community on, 46–47
facilitation of modding, 114
use of as positive, 68–69
Internet cafés, 23
interpersonal neurobiology, 59
iPod
dominance of, 106
increased sales of and podcasting, 128
. *see also* digital media players
isolation and use of MCTs, 68–69

Jacobs, Jane, 17
Janoff, Dean, 200
Johnson, Calvin, 18
Jones, Calvert, 227
journalism
before advent of mobile phones, 216–17
disadvantages of wireless devices, 222–23
disaster coverage, 218–19
editors' use of MCTs, 221–22
and fragility of MCTs, 221
importance of mobile phones, 216
news sources and wireless communication, 217
photographers, 219–21
text communication, importance of, 222
walkie-talkies, 216–17
wire services, 219

Katz, James E., 142
Kenardy, Justin, 193
keyboard, 152
Klein, Britt, 193
Kopomaa, Timo, 12
Kuyper, Jim, 146

La Follette, Robert, 28
land, importance of, 46
Lange, Alfred, 193
laptop computers, 9
Licklider, J. C. R., 108
loneliness and use of Internet, 68–69
Luce, Kristine, 191

Machines That Become Us (Katz), 142
Maheu, Marlene, 198
Maine, e-waste disposal program in, 81–82
malware, 96–97
managed care companies and e-therapy, 196
markets, 160
Martin, Alex, 218–19, 222

Maslow, Abraham, 144
Massachusetts, e-waste disposal program in, 81–82
Massachusetts E-Health Collaborative, 186
Massachusetts General Hospital, policy on use of MCTs, 182
The Matrix (film), 152
McCafferty, Kelly, 193
McLuhan, Marshall, 47, 107, 109
MCTs (mobile communication technologies)
  actual use of *vs.* availability of, 170
  addiction to, 50, 230–321
  Al Qaeda's use of, 170
  and backstage negotiation, 166–68
  and collective action, 163–65
  confusion, 169
  and disaster recovery, 164–65
  and emotional attachments, 60–62
  fragility of, 221
  importance of in developing nations, 165
  importance to networks, 162–68
  isolation and use of, 68–69
  and journalism (*see* journalism)
  and location of expertise, 165–66
  low credibility of, 169
  misuse of (*see* cyber-crime)
  psychological impact of, 47
  reduction of nonverbal expression, 71
  and relationships, 227–28
  security risks of, 169–70
  and work overload, 168
ME++: *The Cyborg Self and the Networked City* (Mitchell), 18
media
  advocates of "marketplace rules," 123
  lack of alternatives, 123
  and power, 152–53
  public (*see* public media)
  . *see also* podcasts
media environments, use of media technology to alter, 107
media of communication, multifacetedness of, 9–10
medicine. *see* healthcare
Meloy, J. Reid, 100
Menino, Thomas, 32, 33
mental health. *see* e-therapy
Messaris, Paul, 143
meta-control, 108, 110–12, 118
Metro/Electro Polis, 62
"The Metropolis and Mental Life" (Simmel), 44
metropolis as switchboard, 8
Meyrowitz, Joshua, 107, 109
microcoordination, 163–64, 169
Microsoft Zune, 111–12
Miller, Thomas, 200
misuse of ICTs, 93
  . *see also* cyber-crime
Mitchell, W. J., 17–18
mobile ICTs. *see* MCTs
mobile mediapolis, 49
mobile phones
  and absent presence, 50
  addiction to, 50
  adoption of, 227
  advertising (*see* advertising)
  Al Qaeda's use of, 170
  ban of in New York City schools, 15
  as bodies, 50
  and cyber-fraud, 99
  and cyber-violence, 101
  and dissolution of boundaries, 43
  as download devices, 152, 153
  driving with, 230
  as entertainment centers, 153, 154
  as fashion, 47, 148
  in healthcare setting, 178–79
  impact on cities, 43
  importance of in developing nations, 165
  interference with medical equipment, 180–82
  and isolation from reality or community, 153, 154
  and journalism (*see* journalism)
  as keyboard, 152
  lack of study of, 40
  malware attacks on, 97
  and microcoordination, 163–64
  multifacetedness of, 9–10
  multifunctionality of, 226
  as necessary, 144

as place, 12, 49
Razr, 110–11
sales of, 9
as self-expression, 148–50
and sense of urban place, 48
smart phones, 10
targeting of by hackers, 96
texts of, 143
time, ability to manipulate, 151
ubiquity of, 40, 141, 142
use of in Philippine protests, 163–64, 169
vocal aggression, 14
mobility
of humans, 7–8
of ICTs, 61–62
modding
challenges of research on, 118–19, 120
collaboration, 116, 118
and place, 119
of POD devices, 115–18
popularity among gamers, 114–15
use of Internet, 114
modding communities
goals of, 114
influence of, 113
Rockbox, 115–18
study of, 113
modifications of technology, 107, 109
. *see also* modding
monopolies of telephone companies, 30
Morawska, Alina, 190
"more" in advertising, 150–51
multifacetedness of communication devices, 9–10
multitasking, 227, 230
muni Wi-Fi
cities using, 31–33
definition of, 23
and economic development, 25
future of, 34–35
government intervention into market as unfair competition, 29–30
impact on cities/communities, 35
legal action on, 31
locus of government authority, 26–27
models of, 31–34
opposition to, 26–27, 29
participants in, 25–26, 31–33
reasons for providing, 24–25
regulation of, 26–27
support for, 30
Murdoch, Janice, 190–91, 198

Napster, 98
National Health Information Network (NHIN), 186
National Public Radio (NPR), 129–30, 131
nature theme in mobile phone ads, 145
neighborhoods, mixed-use, 17
networked organizations
actual use of MCTs, 170
advantages of, 160–61
"anytime, any place" connectivity, 227
boundaries, relaxing of, 164–65
and expertise, 165–66
importance of ICTs/MCTs, 161, 162–68
microcoordination, 163–64
Powell's model for, 160–61
risks and limitations of MCTs, 168–70
security risks of MCTs, 169–70
virtual teamwork, 165
neurobiology, 59, 65–67
New Millennium Research Council (2005), 26
New Urbanists, 17
non-renewable materials, use of in electronic devices, 79
nonsustainable society, 78, 87
nonverbal communication
importance of, 59, 69–71
lack of in cyber-stalking, 100, 101
reduction of by MCTs, 71
Norwood, Stacey, 222
No Sense of Place (Meyrowitz), 107
NPR (National Public Radio), 129–30, 131

OECD (Organisation for Economic Co-operation and Development), 78
Oldenburg, Ray, 12
Olsen, Richard, 227
Online Anxiety Prevention Program, 193

Organisation for Economic Co-operation and Development (OECD), 78
organization, hierarchical, 160
organizations, networked. *see* networked organizations
Osborne, Megan, 191
Oudshoorn, Nelly, 107

pagers, 177–78
Palser, Barb, 131
patient safety, 185
. *see also* healthcare
PBS, 130–31
pedophiles' use of mobile phones, 101
People Power II demonstrations, 163–64, 169
Pépin, Michel, 192
personal space, 14
Pettersson, Richard, 192
pharming, 98
Philadelphia, 31–32
Philippines, protests in, 163–64, 169
phishing, 98
phones, mobile. *see* mobile phones
photographers and wireless MCTs, 219–21
physical co-regulation, 67–68
physical proximity
and health, 59
and resonance, 72
Pierce, Dann, 143
Pierce, John, 190
Pinch, Trevor, 107
place
city as, 42
disconnection of person from, 8, 10–11
displacement theories of, 44
and identity, 48
manipulation of, 106
and meta-control, 110–12
mobile phones as, 12, 49
and modding, 119
Oldenburg on, 12
poor in, 42
place-based experience, disengagement from, 229
podcasts
acceptance of, 128
as alternative media, 124
audience of, 132, 136
characteristics of, 132
as civic involvement, 136–37
community, 132–34
and community, 138
and democratic pluralism, 124, 136, 138
description of, 127–28
as form of public media, 124–25
and increased sales of iPods, 128
by individuals, 132–33
as indulgent, 132
lack of dialogue in, 134, 137–38
mobility of, 124
most popular, 133*f*, 135
new terms related to, 127
as public space for debate, 138
specialty, 133*f*
study of, 124–26
targeted to niche audiences, 132
POD devices
digital media players, 106
modification of, 105–6, 115–18
reconfiguration of, 108
Rockbox, 115–18
user interfaces, 106
portable on-demand devices. *see* POD devices
portable public privacy, 13
Postel, Marloes, 192
Postman, Neil, 154
post-traumatic stress disorder, 193–94
Powell, W. W., 160
power, 152–53
presence, psychological, 12, 15
privacy, portable public, 13
private sociability, 44
psychological disconnection in public spaces, 15
public broadcasting
in Canada, 129, 130, 131
Canadian Broadcasting Corporation (CBC), 129
dissatisfaction with, 131

original intentions of, 126, 129
PBS, 130–31
public radio, 129 (*see also* NPR)
in United States, 129–31
Public Broadcasting Act of 1967, 129
public broadcasting system (PBS), 130–31
public media
audience of, 136
changes in definition of, 137
podcasts as form of, 124–25
shift in values of, 136
. *see also* public broadcasting
public radio, 129
. *see also* NPR
public space
and boundaries of publicness and privateness, 14–15
designing of, 19
impact of new media, 43
privatization of, 12, 13, 44, 48–49
psychological disconnection in, 15
Public Television (Carnegie Commission), 129
Publilius Syrus, 230

quality theme in mobile phone ads, 144–45

radio
decline of listeners, 106
models of, 124
radio, public. *see* public radio
radiology, use of telementoring in, 177
rape, virtual, 101
Razr mobile phone, 110–11
recycling of e-waste, 78–79, 80, 81, 82, 83
regulation
of Internet, 102
of telecommunication, 26–27
of utilities, 27–29
relapse prevention, 191
relationships
bias toward face-to-face communication, 61–62
and MCTs, 227–28
religious practices, mobile communication in, 50
replacement, 12
resonance, 72
Resource Conservation and Recovery Act (RCRA), 81
Restriction of Hazardous Substances (RoHS) directive, 84
rhetorical vision, 144, 146, 150–53
Rholes, W. Steven, 66
Richards, Jeffrey, 193
ringback tones as self-expression, 149
Ritterband, Lee, 192
Rockbox project, 115–18
Rogers, Everett, 141
RoHS (Restriction of Hazardous Substances) directive, 84
Rosa, Virginia, 193
Rosenfield, Maxine, 190
Rotzoll, Kim, 142
Rushkoff, Douglas, 152
Rwanda, 165
Ryan, Frances M., 28

safety and security theme in mobile phone ads, 146–47
Sahr, Robert, 26
Sanders, Matthew, 190
San Francisco, 32
scams, 99
Schafer, R. Murray, 13–14
Schoenholtz-Read, Judith, 200
Schopp, Laura, 196, 197
Seeman, Bob, 198
Seeman, Mary, 198
self-expression as theme in mobile phone ads, 148–50
September 11, 2001, 221
shared body state, 70
shyness and use of Internet, 68–69
Siegel, Daniel, 59, 66, 69, 70, 72
Simmel, Georg, 44, 52
Simpson, Jeffry A., 66
Skinner, Adrian, 196
Slashdot, 114

Slingbox, 112
smart cards, 184
smart mobs, 164
smart phones, 10
Smillie, Evelyn, 190
smishing, 99
Smith, Jennifer, 217
SNOMED (Systematized Nomenclature of Medicine), 182–83
society, technology's influence on, 107–8
solid waste disposal, 80
sound, 13–14
South-East Asian Earthquake and Tsunami Blog, 170
Southeast Asia tsunami (2004), 164–65, 170
space
  disconnection from, 50–51
  mobile phone's ability to manipulate, 151
space, acoustical, 13–14
space, material, 47
space, personal, 14
space, physical, 12
space, public. *see* public space
Stenberg, Daniel, 116
Strom, Lasse, 192, 196
Student Bodies, 194–95
sustainability, 77–78
swarming, 164
switchboard, metropolis as, 8
Systematized Nomenclature of Medicine (SNOMED), 182–83

Tate, Deborah, 192
Taxpayer Relief Act, 82
Taylor, Robert, 108
teaching and use of technology in classroom, 208–14
teamwork, virtual, 165
technology
  adoption of, 141
  collaborative development of, 108–9, 116, 118
  early adopters of, 113–14
  influence on society, 107–8
  modification of (*see* modding)
  responses to, 141
  restrictions on user control, 111–12
  as self-expression, 148–49
  users' influence on, 108
technology theme in mobile phone ads, 145
Tedeschi, Gary, 190
telecommunication, regulation of, 26–27
Telecommunications Act of 1996, 123
telementoring, 177
telepathy, 50
television, user control of, 112
text messaging, low credibility of, 169
texts, of mobile phones, 143
therapy and ICTs. *see* e-therapy
"third places," 52
Thomas, Katie, 219
time, mobile phone's ability to manipulate, 151
Times-Picayune (New Orleans), 221
touching, fear of, 51
transcendence as theme in mobile phone ads, 150–53
Trojan horses, 97, 98
tsunami (2004), 164–65, 170
The Tuning of the World (Schafer), 13–14

Unbox, 111
United States
  e-waste regulations, 81–83
  public broadcasting in, 129–30, 131
  regulation of Internet, 102
United States Government Accountability Office, 79
universal waste, 81
utilities, 27–29
utopian appeal of mobile phone ads, 151

Vargo Daggett, Becca, 32
Verizon Wireless, 85
video game players, popularity of modding among, 114–15
viruses, 96
visual cues and emotional response, 71–72
vocal aggression, 14
VoIP (voice over Internet protocol), 179–80

Walkman, 50, 229
Wall, David, 93–94, 98
Wallace, Patricia, 227
Waste Electrical and Electronic Equipment (WEEE) directive, 83–84
weak ties, 164
Webber, Melvin M., 8
WEEE (Waste Electrical and Electronic Equipment) directive, 83–84
WIAT-TV (Birmingham, Alabama), 222
Wi-Fi
  definition of, 21–22
  impact on cities, 43
Wi-Fi, municipal. *see* muni Wi-Fi
Wi-Max, 22
Winer, Dave, 127
Winett, Richard, 192
Wing, Rena, 192
Winzelberg, Andrew, 191, 194
wireless communication and journalism. *see* journalism
wireless fidelity. *see* Wi-Fi
wireless local area networks (WLANs), 179
Wireless Philadelphia, 31–32
Witherspoon, Gary, 217
WLANs (wireless local area networks), 179
Wood, Jennifer, 200
working models of attachment, 64, 66–67
work/life balance, 168, 227
worms, 96

XML (Extensible Markup Language), 183

Yager, Joel, 191
Yu, LiAnne, 148

Zabinski, Marion, 191, 194–95
Zack, Jason, 196
Zhu, Shu-Hong, 190
zoning, 16–18
Zune, 111–12